Sustainability

A Way to Abundance

Sustainability
A Way to Abundance

Corrado Sommariva

CRC Press
Taylor & Francis Group
Boca Raton London New York

CRC Press is an imprint of the
Taylor & Francis Group, an **informa** business

CRC Press
Taylor & Francis Group
6000 Broken Sound Parkway NW, Suite 300
Boca Raton, FL 33487-2742

First issued in paperback 2020

© 2018 by Taylor & Francis Group, LLC
CRC Press is an imprint of Taylor & Francis Group, an Informa business

No claim to original U.S. Government works

ISBN-13: 978-0-8153-6168-8 (hbk)
ISBN-13: 978-0-367-60721-0 (pbk)

Visit the Taylor & Francis Web site at
http://www.taylorandfrancis.com

and the CRC Press Web site at
http://www.crcpress.com

This book is dedicated to my family: they are the base and the inspiration of my life.

"We have to look without fear to the technological transformation of economy."

Papa Francesco

Contents

What If?

WHAT IF THERE WAS more to sustainability than what we think now? What if the pristine beauty of nature could be restored by humans in the process of their evolution and could be harmonized with modern generations' way of living? What if there was a universe of abundance with which humans have been endowed?

What if water was a key to that abundance?

Foreword

MOST OF US OWE our high standard of living to the technological innovations developed during the last century—affordable energy, rapid transportation, and advanced information systems. However, most of us fail to realize that we have inherited, along with our good fortune, a bevy of environmental and social problems.

Sustainability is an elusive concept, and many have tried to define the idea of what it is. But rather than trying to frame the concept, this book develops a supporting structure, a "dynamic sustainability," that follows a natural set of steps to reach an established goal, which is the "state of increased energy and abundance." This book provides the tools to navigate this "road to a better future" by explaining the thermodynamic approach to the concept of sustainability, giving ideas, proposing methods and formulae, and suggesting actions. The book vehemently supports the idea that real change comes when industrial processes are designed to be more economically, socially, and ecologically beneficial rather than merely less polluting.

This book is intended for people from all walks of life. Primarily, decision makers, stakeholders, students, and the public in general can reap benefits from the book by blending the use of thermodynamic sustainability concepts, tools, and ideas with their current projects in order to best attain a sustainable environment.

Dr. Rashid Al Leem
Chairman
Sharjah Electricity and Water Authority

Preface

As a former municipal politician and political analyst who specializes in sustainability, stakeholder awareness, and environmental diplomacy, it is often difficult to find reasons to be optimistic. Climate change is addressed at elaborate international events that conclude with self-congratulatory speeches aimed more at ensuring the speakers' presence in the following day's headlines rather than establishing a definitive strategy to combat the global environmental challenges with which we, as a society, are faced. Concepts come, and concepts go. Keywords are repeated *ad nauseam* or indeed are positively abused by policymakers and elected representatives whose understanding of complex issues is often reduced to a summary prepared by an overworked aide. Workshops, seminars, and conferences blossom on the pages of the agendas of students, researchers, scientists, industrial professionals, managers, civil servants, and civic leaders. Entities compete with each other in the organization of annual encounters attended by a traveling circus of experts who purport to already understand the issues and who are, more likely than not, seeking by their presence to attract attention to their own current methodological advances rather than to listen and learn from fellow participants.

Human ambition, political dogma, ideological prejudice, and professional exclusivity often prove to be important obstacles to a broad social dialogue concerning the principal challenges of the twenty-first century, among which the environment must take precedence, and yet, it is this social dialogue that our world requires. Information available to all stakeholders within what is now described as the *Quintuple Helix* leads to awareness and, more importantly, concern. Concern results in engagement and the broader the foundations of such involvement the greater the social consensus created. It is this concord that can guarantee the political and consequently the economic continuity necessary to effectively redress the environmental conditions which now exist. Therefore, we live in an age in which theorists must not only think, create, or innovate. They must communicate to those who exist beyond their comfort zone. They must transmit not only their thoughts but emit clear, practicable notions that society as a whole can embrace. Corrado Sommariva has done just this.

Sustainability: A Way to Abundance is the work of an engineer who, by writing this book, has demonstrated the personal conviction and determination that a new, highly positive approach to the concept of sustainability is not only possible from both a thermodynamic and an environmental perspective, but that we are capable of creating a new wealth, whereby the damage the biosphere has suffered due to nearly 200 years of anthropogenic activity can be repaired while permitting future generations to enjoy the full fruits of mankind's technological advances. By directly questioning the traditional premise that sustainability is the preservation of the status quo and the conservation of the planet's resources for the benefit of our children, Sommariva introduces the reader to a future of biocentrism, in which advances regarding exergy, as opposed to energy; the mining of recoverable technospherical prime materials such as mercury, lead, and iron; and the restoration of water to its rightful place as the prime source of life itself leads to a potential situation of abundance, whereby a restored, pristine natural environment is possible. The debate, after having read Sommariva's clear, logical discourse is no longer *"Are we doing enough to counteract the negative ecological panorama with which we are faced?"* If one accepts Sommariva's premises, we are now capable of entering into a revolutionary phase of *resource optimism*, a tendency to accumulate that would not only supply modern society with its energy requirements but would permit our natural surroundings to recover their original splendor and equilibrium.

The opportunity to establish a more intimate and diligent relationship with our surroundings, while not having to deny future generations the technological comfort of our age, has profound social, economic, and political repercussions. *Sustainability: A Way to Abundance* will, as do all important theses, provoke discussion and debate. It is the stated reflection of careful consideration by an expert who has proved himself capable of moving beyond the frontiers of his own professional community to interact with other, equally important social sectors. The vision that this book describes is a necessary proposal to be presented to the global forum of environmental technology and policy. Sommariva has taken the first and most crucial step by contemplating an approach in which a wide range of actors should be involved. It is now for those people, be they experts or hitherto uninformed members of our community, to reciprocate by reading the work of Sommariva with all due academic thoroughness and, perhaps more importantly, with human optimism. *Sustainability: A Way to Abundance* translates technical theory into social discourse. It becomes, therefore, an informed but extremely accessible source of social consensus and political continuity, two vital elements if as a global community we are to progress.

Richard Elelman
Head of Public Administrations of EURECAT-CTM

Introduction

THE CONCEPT OF SUSTAINABILITY came relatively late in the water business, in which I have been working for years. It was introduced along with the industry's endeavors to renew itself, adapting to more modern exigencies of human life and, as time passes, fulfilling a promise of social commitment toward the environment, energy conservation, and sustainability. A great emphasis today is placed on the sustainability of energy solutions applied to the water business; as the water sector got more and more engaged in sustainability, my journey into this world started.

When my company and I were working in the largest photovoltaic plant in the Gulf Corporation Council (GCC) at His Highness Mohammed bin Rashid Al Maktoum Solar Park, and later, when I became involved in the first unique program for renewable desalination generation in 2012, which was launched by MASDAR in Ghantoot, sustainability became an essential component in my profession.

As I was delving into this discipline, which was completely new and somehow unfamiliar to me, I began to realize that water is essential to the overall goal of designing a sustainable future for the next generations, and without drastically changing our approach to technology, the overall goal of sustainability may not be realistic.

But indeed, the words of a spiritual leader, Dr. Wayne Dyer, prompted me to take the plunge and argue the concept that there is abundance beyond sustainability from an engineering point of view These stirring, wonderful words were: "See the world as an abundant, providing, friendly place. . . . Remember that no one lacks abundance. The supply is unlimited. The more you partake of the universal generosity, the more you'll have to share with others."

I am aware that as inspiring as it sounds, this statement may seem paradoxical considering today's reality of millions of people in poverty and hunger and the increasing gap between poverty and wealth. However, this does not mean that our approach to economy and business should prevent us from aiming at achieving abundance.

It was when I realized that the idea of sustainability itself, in the way that we define it today as a limitation to what we can provide to our planet, that I decided to write this book. It aims, therefore, to demonstrate that starting from the abundance of renewable energy, abundance of resources can be continued all the way downstream in society and the economy.

As of today, the most popular meaning of sustainability was given in the Brundtland Report of 1987; it defines *sustainable development* as development that meets the needs of the present without compromising the ability of future generations to meet their own needs (United Nations, 1987), but I strongly believe that there is more to sustainability.

There is a universe of abundance that is beyond our current definition of sustainability, and in which technology plays a key role. The traditional approach to sustainability is based on the preservation of the status quo, whereby our responsibility toward future generations is simply to conserve planet resources to enable life as we know today.

On the other hand, thanks to the constant energy flow that our planet receives regularly from the cosmos, our remittance to future generation can be much more comprehensive. It can include building up more and more abundant resources and achieving a positive energy balance aiming at storing rather than abstracting energy from the Planet. Furthermore, an economy can be based on recovering the degradation caused to the environment and creating beauty, which can enable future generations to live with standards toward their own lives and the environment, which are not conceivable today.

In this journey, taking us from sustainability to abundance, water is the key element because of its capacity to create life and to enable harvesting the energy that is irradiated to our planet every day from the cosmos.

It took quite a while for me to resolve to publish this book. I have been wondering if anyone would read it, and if the concept that I propose here could be comprehended. For a long time, I thought that this book could be seen as a clumsy attempt to bring a responsible approach to environment as part of the human evolution in today's industry and market and link it to the concept of sustainability. What scared me most was that I am venturing outside my comfort zone as an industry professional in the water and power market.

My previous publications were written mainly for engineers. When I coauthored the book *Water, Energy, and Food Sustainability in the Middle East: The Sustainability Triangle,* the approach was mainly from an engineering point of view—a specialist's take on this subject. Instead, this book is addressed to a much broader audience: students, scientists, industry professionals, researchers, and even economists and managers. My intent is to approach sustainability from both a theoretical and practical point of view, for a readership that is not familiar with sustainability concepts. I also hope this book will be useful for policy planners.

The main driver that prompted me to write this book was my desire to impart some new and innovative ideas on the subject of sustainability. I aimed to use this book to provide a new approach to sustainability—one that would be able to make a difference in the way that we understand and perceive our relationship with nature, the environment, and the energy management of the planet, which drives a hope for achieving abundance. I have tried to demonstrate that this hope is possible from both a thermodynamic and an environmental point of view.

I hope that this book will help to inspire readers to commit to the ambitious task of restoring the planet to its original beauty, and help them understand the key role that water plays in our endeavor to fulfil this objective. The book supports technology and human social and civil progress as perhaps the only remaining available tools to drive this positive change.

The analysis proposed in this book is centered on a thermodynamic approach to sustainability and energy. It aims at providing a stronger cornerstone on the definition of the peculiar correlation between energy (exergy), environmental impact, and sustainability.

The approach to sustainability presented in this book is not a way to bestow upon the next generation something that does not endanger their future needs. However, I believe the purpose of this text is much more challenging and motivating: to regenerate the wealth and abundance with which previous generations have been endowed, and then build more.

Going through the Book

THE BOOK STARTS WITH a simple thermodynamic approach to the concept of sustainability and the definition of the concept of a dynamic sustainability, which is driven by the constant energy flux coming from the sun. This energy flow, in absence of relevant anthropogenic activities, drives the planet towards a state of increased energy and abundance of resources.

The state of pollution, unsustainability, and environmental disruption is, therefore, linked to a decreased energetic state of the planet through the chemical energy associated to the state of formation of the elements composing the biosphere. After an analysis of the energy approach to sustainability, the book describes the concept of energy sustainability and abundance applied to resources of the planet and how a natural abundant cycle of resources accumulation in nature is altered by anthropogenic activities.

The third section of the book illustrates also how the correct use of technology will be capable to respond to mankind's demand for a more affluent life and include plenty of access to nature as a key criterion for this affluence and for the reinstatement of the Earth energy abundance. In particular, the section describes how, taking advantage of the abundance in renewable energy, abundance of resources can be flowed down in all aspects of human life.

The topics presented in this third section of the book start with an initial screening of the environmental theories and the relationships among environmental impact and sustainability with technology affluence.

After a critical analysis on how these theories should be relooked at based on today's lifestyles, a new proposition of environmental impact is proposed where the role of technology and the relationships between environmental impact, affluence, and technology are reversed, and the role of technology in mitigating and even reversing environmental impact is justified.

The fourth section of the book describes the essential economics of sustainability and demonstrates the economic feasibility of the new wave of economic development and how it is possible to develop a growing economy that is not at odds with sustainability and with the preservation and reinstatement of environment.

The last section is related to water and its key role in ensuring abundance and how the cradle-to-cradle approach should be applied to new water schemes to preserve and augment our biosphere.

To perhaps the disappointment of many readers, the book contains many formulas. I ask for the reader's understanding in this respect; formulas are a component of my engineering background, but in this case, they are also a necessity. In fact, I wanted to introduce the concept of abundance and the relationship between energy and sustainability in a form that would be arguable and demonstrable also from a purely theoretical and numerical point of view.

On the other hand, since the book is directed in practice to everyone interested in sustainable development, I tried to translate these formulas into accessible language so that the book may look like a hybrid between a technical manual, a research and development essay, and a seminal book on sustainability.

I understand this may look strange to many readers, and I kindly ask their understanding. On the other hand, I believe it is essential to the goal of sustainability that all sectors are involved, and the sustainability aspects are captured from as many point of views as possible.

Acknowledgments

I HAVE TAKEN GREAT INSPIRATION in writing this book by attending the series of conferences held by Alleem Business Congress and by my personal relationship with His Excellency Dr. Rashid al Leem, whom I would like to thank for his visionary approach to sustainability and for his commitment towards a sustainable future.

I also would like to acknowledge my colleague engineer Rosario Blasetti who had the patience to read through the book and the formulas, providing some useful insights, and my colleague engineer Onnis, who helped with illustrations. Great thanks to Professor Capannelli, who guided me through this publication, and Dr. Richard Elelman of Eurecat for his strong inspiring introduction.

Last but not least, I am also grateful to my company, ILF, for the support given in the preparation of the book and the commitment to a sustainable future in the daily business practice.

Author

Dr. Corrado Sommariva is the managing director and member of the board of ILF Consulting Engineers Middle East.

Dr. Sommariva has a PhD in chemical engineering from Genoa University and a diploma in management from Leicester University.

In November 2013, Dr. Sommariva was conferred an honorary doctorate by Herriot Watt University in recognition to his contributions to chemical engineering.

In May 2014, Dr. Sommariva was conferred by the president of Italy the title of "master of work" along with the cross of merit for his distinguished contribution and career.

In 2017, Dr. Sommariva was listed among the 50 most impactful smart cities leaders and among the 25 most impactful water leaders in the world.

Dr. Sommariva is an honorary professor at Genoa and L'Aquila Universities where he holds regular courses on water treatment, energy efficiency, desalination, and economics. He has published over 80 papers on various leading-edge technical research and economics and has published two books on water management and economics.

Dr. Sommariva co-authored the book *Water, Energy & Food Sustainability in the Middle East: The Sustainability Triangle*.

Dr. Sommariva was president of the International Desalination Association for the term 2011–2013, served as co-chairman of the Technical Program for the 2009 IDA World Congress in Dubai, and was the chairman of IDA's recently held conference, "Desalination Industry Action for Good," a landmark event held in Portofino, Italy that focused on water and social responsibility.

Along with his company, ILF, Dr. Sommariva was engaged in the development of the largest photovoltaic project in the GCC at His Highness Mohammed bin Rashid Al Maktoum Solar Park, and he was one of the promoters of the unique program for sustainable desalination that was launched by MASDAR in Ghantoot.

Dr. Sommariva has been president of the European Desalination Society (EDS) and chairman of the World Health Organization's committee for the establishment of safe drinking water.

CHAPTER 1

Energy and Sustainability

1.1 A NATURAL CYCLE OF ENERGY ABUNDANCE

The leading point in this part of the book is related to the Earth energy level and the balance of energy in the Earth as an indicator of the sustainability.

It is relatively common knowledge that the Earth receives abundant energy from the sun and that this energy is, in theory, more than sufficient to support today's and tomorrow's projected energy requirements for anthropogenic activities.

This is surely good news. Undoubtedly, sooner or later technologies will be developed, and these will be capable to harvest as much clean energy as it is required for humanity from renewable sources at affordable cost. Sooner or later, technology will be reaching a point where this renewable energy will be flowed down to all sectors of human activity: transportation, manufacturing, etc. Eventually we will cease the dependence upon fossil fuel.

While that will be a great achievement for humanity, the issue of energy and sustainability will not be completely solved.

It is not commonly understood, in fact, that the energy (it will be later defined as exergy) of the Earth is strictly associated with our biosphere and to the chemical state of the elements that compose the planet. In the biosphere, the elements that are composing the Earth are come together in different chemical and biological forms, and these forms are the results of endothermic reactions that are occurring thanks to the sun energy.

Biosphere can be defined as the system comprising all living organisms on the earth with the exception of modern man: Plants and animals consequently belong to the biosphere (Smil V., 2017).

Modern man and all anthropogenic products belong to a system that can be called the techno sphere; therefore, buildings and machines belong to the technosphere.

In more technical terms, the exergy of the Earth is associated with the chemical potential of the elements that compose the biosphere and the techno sphere: the lower their chemical potential and the higher their level of oxidation the lower the energy of the Earth.

This chapter of the book proposes a link between sustainability and the exergy level of the Earth. In particular, non-sustainable practices lead to a decrease of the exergy of the Earth (Figure 1.1) and eventually to the exergetic death of the planet, while sustainable practices would lead to a gradual increase in the Earth energy, to a state of abundance.

The sustainability challenge is therefore related not only to the elimination of fossil fuel dependence but also to the recovery and management of the exergy levels in the biosphere. Unfortunately, these have been gradually reduced because of the pollution caused by anthropogenic resources and because of the gradual shrinking of the biosphere volume.

The energy of the Earth is composed both by the energy associated with the fossil fuels and other non-renewable sources in the lithosphere, but also by the energy associated with the chemical potential of elements such as carbon and nitrogen.

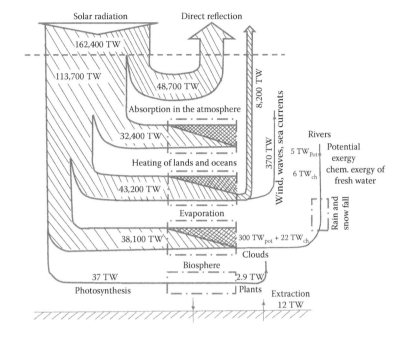

FIGURE 1.1 Distribution of the exergy flows above the earth surface. (From Szargut, J. T. 2003. *Energy.* 28(11): 1047–1054.)

In particular, when carbon or nitrogen, for instance, are present in form of biomass, soil organic carbon, or ammonia fixed in the soil, their chemical potential and therefore their energy is higher than when these components are in the CO_2 or free nitrogen form in atmosphere.

In the same manner, the energy of the Earth is associated with the energy of other elements such as mercury, lead, iron, etc. recovered and stored in the technosphere in the form of accessible and further usable prime material or technical nutrients, (McDonough et al., 2013) in opposition to being spread as waste in meadows woods, rivers, and ecosystems linked to the natural humanity cycle.

This makes the sustainability challenge slightly more complex, but it offers an enormous opportunity: this opportunity is to tap in into the sun's renewable energy abundance to create other abundance all the way downstream. The opportunity that is offered is to use this energy to satisfy anthropogenic demand, restoring the biosphere and, in a second stage, to increase the exergy level of the planet, creating an energy and resource store for humanity well-being and for future generations.

1.2 RETURN TO NATURE ENERGY ABUNDANCE AND BUILD MORE

The day-to-day phenomena occurring on the planet as we know today require energy: energy is required for waves, wind, water evaporation, rain, snow, and for all forms of life existing in nature (Szargut, 2003).

Solar radiation is the main source of energy sustaining all natural processes occurring on Earth and providing the energy for the regeneration of the planet life.

There are different terminologies for energy, but from now onwards, the term exergy will be used instead of energy. The definition of exergy was then introduced for the first time by Rant (Rant, 1956) as a thermodynamic function measuring the maximum work extractable from a system when it is brought at thermodynamic equilibrium with its reference environment.

In simple terms, exergy corrects the fact that, for instance, 1 kJ of electric energy has much more value in terms of work that can be extracted from it than 1 kJ of heat.

It also corrects the fact that at 400 °C 1 kJ of heat has a much higher value than at 40 °C.

Basically, this concept has been known since 1824 when Carnot (1824) gave his definition of the second law of thermodynamics, stating that the extractable work from a heat engine is proportional to temperature difference. The exergy concept became known (Carnot et al., 1824) after that when Gibbs (1873) introduced the concept of available work, corresponding exactly to the present definition of exergy.

Regarding solar energy, it is possible to consider solar radiation as coming from an infinite energy storage interacting with the local environment. In this case, exergy can be defined as the maximum amount of work that can be extracted from a system in the process of reaching the equilibrium with the local surrounding environment.

In basic terms, exergy is a measure of quality and order of the system which is referred to; the higher the exergy, the higher the usable energy of a system.

The majority of all anthropogenic processes today require energy (exergy), which is provided by the consumption of the fossil fuel resources that have accumulated on the planet for millions of years.

1.3 THE EXERGETIC FLOWS OF THE EARTH

Several studies were carried out aiming at analyzing and computing the exergy consumption of anthropogenic activities (Szargut, 2001) and aiming at defining a sustainable exergy consumption.

Figure 1.1 (Szargut, 2003) illustrates schematically the flux of exergy in the Earth in the current era and, therefore, in the presence of energy consumption caused by anthropogenic activities.

As it can be seen from the figure, the mean exergy flow of solar radiation reaching the external layers of the amounts to 162,400 TW.

A reflective phenomenon of the solar radiation from the upper layers of the atmosphere (48,700 TW) occurs along with a dissipative process of some part of solar exergy in the atmosphere (32,400 TW). This is caused by the absorption of solar radiation and re-emission of the infrared radiation heating the land and oceans, which amounts to 43,200 TW.

A prevailing part of this radiation is transformed into the infrared radiation of the Earth's surface, partially transmitted to the cosmic space and partially absorbed by the atmosphere and re-emitted to the Earth's surface. This is currently causing the greenhouse effect.

Only a small part of the radiation heating the Earth's surface (about 370 TW) is transformed into the mechanical exergy of wind, waves, and sea currents. This represents a typically dissipative process.

Among other data, it is interesting that Figure 1.1 shows also the flow of exergy in the biosphere.

The exergy flow used for the evaporation of water is about 38,100 TW. A small part of this exergy is transformed into a potential exergy of water droplets or ice particles contained in clouds.

The mass of evaporated water (assuming a mean temperature 25 °C) amounts to 15 × 109 kg/s. Assuming the mean height of clouds is 2,000 m, the potential exergy of the clouds generated in a time unit is about 300 TW. A prevailing part of this exergy is used to generate rain and snowfall.

Only a very small part (5 TW) is transformed into the potential exergy of rivers and lakes.

The droplets of liquid or solid water contained in clouds represents a renewable source of fresh water. Its chemical exergy amounts to about 1.42 kJ/kg, if the mean composition of the seawater (salinity about 0.577 mol/kg H_2O) has been accepted as a zero level of chemical exergy.

Considering the mentioned rate of the evaporated water and the fraction of land on the Earth's surface (0.29), it is possible to calculate the total stream of chemical exergy of fresh water reaching the land with the rain and snow (about 6 TW).

1.4 WHY ABUNDANCE IN NATURE?

While the overall incoming exergy level on Earth is 162,400 TW, the exergy that is absorbed by the biosphere is relatively small: in fact, the radiation absorbed by the vegetation (only the active part) has an energy of about 40 TW (exergy about 37 TW).

Furthermore, only a small part is transformed into the biochemical exergy of plants (about 2.5 TW of energy), and about 1 TW of chemical exergy of plants is consumed by humans. This would appear at odds with the concept of abundance unless the concept of energy or exergy storage is introduced.

A thermodynamic cycle can be considered dissipative when all the energy that it receives is consumed to restore the initial conditions. This is not what, in absence of anthropogenic activities, happens in nature. In nature, a part of the energy provided by the sun is utilized to support life, but another part is converted in chemical exergy, gradually stored, and accumulated. This is a dynamic cycle towards an increased energetic stage.

The conversion of solar energy to chemical energy in the form of substances at higher chemical potential occurs thanks to plants, algae, and some bacteria that contain chlorophyll and, therefore, allows the biosphere to harvest some of the solar energy.

In the process of photosynthesis, the chlorophyll in plant leaves uses solar energy to assemble glucose molecules from water and carbon dioxide. Oxygen is also produced by photosynthesis.

Energy coming as photons from the sun is stored through photochemical reactions in adenosine triphosphate (ATP) and nicotinamide adenine dinucleotide phosphate (NADPH) while producing oxygen from water, as shown in Figure 1.2 (http://book.bionumbers.org/how-much-energy-is-carried-by-photons-used-in-photosynthesis/). This energy is then used to fix inorganic carbon by taking carbon dioxide from the air and transforming it into sugars that are the basis for biomass accumulation and long-term energy storage in the biosphere.

In the photosynthetic process, the energy of the light is absorbed by the chlorophyll molecules of the plants to convert carbon dioxide and water into carbohydrates and oxygen gas. As long as elements such as nitrogen, sulfur, and phosphorous are available, proteins, fat, and other compounds are also synthetized.

Some of the energy is dissipated in cellular respiration whereby carbohydrates proteins and fats are broken down and oxidized to provide energy in the form of ATP for the cell metabolic needs. However, there is a component of energy that is not used and is stored in plant tissues, and there is a phenomenon of decay and littering of the plant tissues that contributes to increase the soil organic carbon.

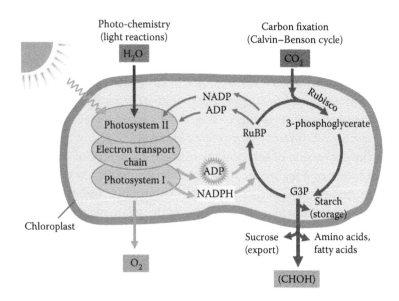

FIGURE 1.2 The flow of energy in the biosphere. (From http://book. bionumbers.org/how-much-energy-is-carried-by-photons-used-in-photosynthesis/)

Exergy of the Earth, therefore, is associated with the exergetic content of its elements and their chemical potential. Carbon in soil or biomass has a chemical potential that is higher than free CO_2; therefore, a loss of these components brings about a loss in the Earth exergetic content.

Photosynthesis has transformed the planet into its present state. While part of the sun's energy is dissipated in the biosphere, some of this carbon is constantly put away in the global store of fossil "fuel" in the lithosphere or accumulated in soil as soil organic carbon at the same time oxygen is emitted to the atmosphere (Wall et al., 2001).

In the biosphere, therefore, both the living and the dead materials have accumulated providing substantial reservoirs of energy in form of chemical energy. As it can be seen from Figure 1.4, in the absence of anthropogenic activities, the carbon balance is positive; it produces and stores organic and mineral carbon compounds more than it uses by oxidizing them. In other terms, the biosphere segregates more CO_2 from the atmosphere than what it emits and uses the related carbon to generate and store more complex and energetically valuable biological components.

Natural processes, acting for billions of years, have built up the deposits in the lithosphere. By this, nature has also created conditions suitable for life on Earth, for example, the availability of oxygen in the atmosphere and the removal of unwanted and toxic substances from the biosphere.

This is basically indicated in the schematic diagram of Figure 1.3.

This is an abundant cycle in the sense that in the absence of the modern man, and therefore in absence of a technosphere, the biosphere is capable not only of upcycling all the natural waste generated by the biosphere renewing life, but it also is capable of putting in store more energy.

This abundance is reflected in the absence of substantial anthropogenic activities in the storage of chemical elements at a higher exergy value resulting from a number of redox reactions in the biosphere.

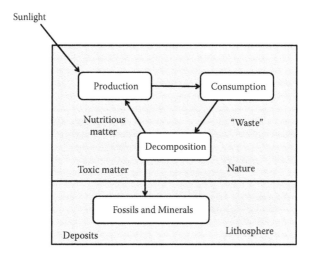

FIGURE 1.3   Resources used by nature. (From Wall, G. et al. 2001. On exergy and sustainable development—part 1: Conditions and concepts. *Exergy, An International Journal.* 1(3): 128–145.)

An example is illustrated by Figure 1.4, which shows the "carbon stocks" in PgC (1 PgC = 1,015 gC) and annual carbon exchange fluxes (in PgC yr⁻¹). Darker numbers and arrows indicate reservoir mass and exchange fluxes estimated for the time prior to the Industrial Era, about 1,750.

As can be seen from Figure 1.4, CO_2 is removed from the atmosphere by plant photosynthesis [gross primary production (GPP), 123 ± 8 PgC yr⁻¹ (Clais et al., 2013)] and carbon fixed into plants is then cycled through plant tissues, litter, and soil carbon and can be released back into the atmosphere by autotrophic (plant) and heterotrophic (soil microbial and animal) respiration and additional disturbance processes

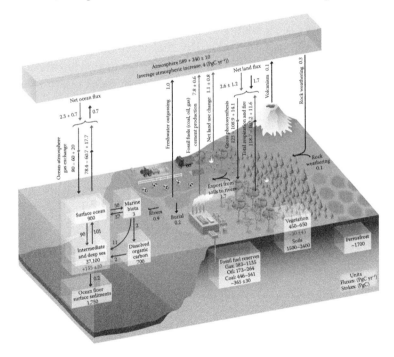

FIGURE 1.4   Simplified schematic of the global carbon cycle. (Ciais, P. et al., *Climate Change 2013: The Physical Science Basis. Contribution of Working Group I to the Fifth Assessment Report of the Intergovernmental Panel on Climate Change.* Cambridge University Press, 465–570.)

(e.g., sporadic fires) on a very wide range of time scales (seconds to millennia). A significant amount of terrestrial carbon (1.7 PgC yr^{-1}) is transported from soils to rivers headstreams. A fraction of this carbon is outgassed as CO_2 by rivers and lakes to the atmosphere, a fraction is buried in freshwater organic sediments, and the remaining amount ($\sim$0.9 PgC yr^{-1}) is delivered by rivers to the coastal ocean as dissolved inorganic carbon, dissolved organic carbon, and particulate organic carbon (Tranvik et al., 2009).

Photosynthesis and respiration are essentially the opposite of one another. Photosynthesis removes CO_2 from the atmosphere and replaces it with O_2. Respiration takes O_2 from the atmosphere and replaces it with CO_2. However, these processes are not in balance. Not all organic matter is oxidized. Some is buried in sedimentary rocks. The result is that, over geologic time, there has been more oxygen put into the atmosphere and carbon dioxide removed by photosynthesis than the reverse.

Table 1.1 shows the main natural processes that remove CO_2, their atmospheric CO_2 adjustment time scales, and main (bio) chemical reactions involved.

Overall, as seen from Figure 1.4, in the absence of sensible anthropogenic activities, the carbon dioxide flux is negative, and, therefore, the endothermic component of the set of reactions prevails, leading to an increase of the energy levels of the planet.

With reference to the set of chemical reactions indicated in Table 1.1, the lower the residence time of carbon in the planet as CO_2, the higher the Earth energy levels.

Clearly, during the past 800,000 years, atmospheric CO_2 has not been always decreasing as the result of biosphere uptake. CO_2 concentration increased from 180 ppm during glacial (cold) up to 300 ppm during interglacial (warm) periods.

In this case while the variations in atmospheric CO_2 from glacial to interglacial periods were caused by decreased ocean carbon storage, these were compensated by increased land carbon storage (300 to 1,000 PgC).

TABLE 1.1 Main Reaction Removing CO_2 from Atmosphere

Processes	Time Scale (Years)	Reactions	Type
	Land Uptake		
Photosynthesis	1–10^2	$6CO_2 + 6H_2O$ sun energy $\rightarrow C_6H_{12}O_6 + 6O_2$	Endothermic
Respiration		$C_6H_{12}O_6 + 6CO_2 + 6H_2O +$ heat	Exothermic
Seawater buffer	10–10^3	$CO_2 + CO_3^{2-} + H_2O \rightarrow 2HCO_3^-$	Exothermic
Reaction with calcium carbonate	10^3–10^4	$CO_2 + CaCO_3 + H_2O \rightarrow 2HCO_3^- + Ca^{2+}$	Exothermic
Silicate weathering	10^4–10^6	$CO_2 + CaSiO_3 \rightarrow CaCO_3 + SiO_2$	Exothermic

Source: From Ciais, P. Climate Change 2013: The Physical Science Basis. Contribution of Working Group I to the Fifth Assessment Report of the Intergovernmental Panel on Climate Change. Cambridge University Press, 465–570.

In other terms, abundance derives from the fact that the biosphere cycles enable the absorption of more energy than what it is required for supporting life and constantly enriches the planet with the products of an endothermic reactions whose energy comes from sun.

1.5 ANTHROPOGENIC IMPACT AND ENERGY DISRUPTION

Below the Earth's surface, a carbon bank consisting of fossil fuels exists. This reservoir has accumulated for millennia, and it is used today as an energy bank to support today's living requirements. Traditionally, the impact of anthropogenic activities in the overall energy balance of the Earth has been associated with the amount of energy withdrawal from this reservoir.

The exergy storage has been possible for millions of years. To a great extent, this occurs also today except that, as indicated in Figure 1.1, anthropogenic activities withdraw from the Earth 12 TW of fossil fuel energy.

For human society, the main source of exergy is fossil and nuclear fuels and the products of photosynthesis comprising food, fuels, and building timber. The sum of the anthropogenic exergy losses has been evaluated by Szargut (Szargut, 2003) at 13 TW, or by Wall and Gong (Wall et al., 2001) at 12 TW. This intensity is in the order of magnitude of 1% of the global terrestrial exergy consumption on Earth.

Though about 340 times smaller than the global exergy loss, this amount lost near the Earth's surface might be compared at first glance with the losses associated with certain processes that occur near the Earth's surface, such as the amount of 37 TW for the photosynthesis associated with the ecosphere and 31 TW for the geothermal effect essential for the landscape. In reality, the exergy disruption generated by the interaction of polluting elements with the biosphere is not accounted in this scheme. This generates a gradual impoverishment of the planet exergy, and, therefore, the comparison is not complete.

Local anthropogenic exergy disruption in highly populated areas is also generally by far greater than that of biosphere exergy accumulation. In this case, the anthropogenic impact may dominate some entire ecosystems.

From the analysis of the energy flow chart indicated in Figure 1.1, the use of 12 TW for anthropogenic activities brings about a gradual decrease of the exergetic state of the Earth.

However, in addition to the fossil fuel reserve, there are two other exergy reservoirs that humanity is currently using: one is the biosphere, and the other one is the exergy level of the technosphere, as schematically indicated in Figure 1.5.

As schematically indicated in Figure 1.5, the technosphere has, with time, accumulated technological products that are the result of several transformations requiring energy. These products, therefore, have an inherent exergetic level associated with their manufacturing state and use, and a non-negligible component of the fossil fuel reserve has been utilized in the conversion of raw materials in manufactured goods.

The importance of these energy reservoirs is completely different, and often it is not completely captured in the whole

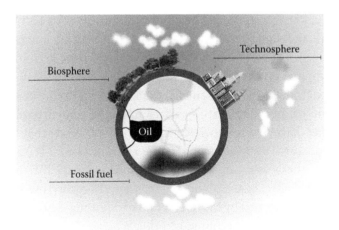

FIGURE 1.5   Energy reservoirs: Fossil fuel, biosphere, technosphere.

sustainability analyses. In fact, when technologies are developed enabling humanity not to depend on fossil fuel, it will be possible to live with a small residual fraction of the existing fossil fuel remaining as a reserve, but it will not be possible to live with a small residual fraction of the biosphere energy reserves.

This would be equivalent to living with a small fraction of organic carbon in soil, of a small remaining fraction of photosynthetic capacity available. This situation would still bring about an irreversible degradation of the energetic levels of the planet that would gradually fill the planet with oxidized reaction products generated by anthropogenic activities.

As the time goes by, the discharge of waste generated by anthropogenic activities on the Earth is interacting with the biosphere more intimately on its exergetic levels than as it was envisaged by Szargut (Szargut, J. T. 2003). Pollution of the biosphere decreases biosphere chemical potential and its energy levels.

Therefore, in addition to the consumption of 12 TW drawn from our fossil fuel resources, it is essential to consider that anthropogenic activities generate waste, solid, liquid, or gaseous, which interacts with the biosphere and further deceases its exergy.

In other words, waste material and waste energy generated by the anthropogenic activities by either burning fossil fuel or municipal or industrial waste leave an ecological mark that is sometime irreversible on the Earth, and it affects sharply the exergetic level of the Earth well beyond the consumption of the stored fossil fuel.

Exergy has been already applied to the study of the fundamental characteristics of ecosystems, with the denomination of eco-exergy. Basically, eco-exergy reflects the chemical exergy and the information stored in the biomass of organisms making up an ecosystem (Susani et al., 2006).

Exergy can be derived also from the second law of the thermodynamics whereby Equation 1.1

$$Ex = (T - T_0) \cdot S - (P - P_0) \cdot V + \sum_{i=0}^{n} (\mu_i - \mu_{i0}) \cdot n_i \qquad (1.1)$$

where U, V, S, and n_i (internal energy, volume, entropy, and number of moles) are the extensive variables of the system; P_0, T_0, and μ (pressure, temperature, and chemical potential) are the intensive variables of the environment; and n_i is the number of moles at related chemical potential.

In this case, for instance, the chemical potential μ of carbon, hydrogen, and oxygen in the form of cellulose or glucose plus oxygen is higher than the chemical potential of carbon dioxide plus water, and this is summed for all the components n_i that are in the biosphere; therefore, in this form, these elements store energy, while in the oxidized form, energy is dissipated.

On the other hand, anthropogenic activities generate pollutants. These are often toxic and affect the overall thermodynamics of the biosphere, as they impact the chemical potential of the elements in the biosphere and affect the exergetic state of the Earth.

This is schematically indicated in Table 1.2 showing few examples of how human pollution can affect the energy of the

TABLE 1.2   Examples of Anthropogenic Pollution and Earth Energy Relation

High Exergy	Low Exergy	Description
		Carbon and oxygen in their oxidized form CO_2 and H_2O have higher chemical potential than cellulose or glucose
		A land before and after deforestation. Deforestation removed the biosphere and therefore no carbon energy stored in its subsoil
		A contaminated land with toxins polluting the biosphere and therefore no capabilities to develop carbon absorption in its subsoil recovered afterwards

Earth through the biochemical potential of the elements that constitute the biosphere.

There could be many more examples than those illustrated in Table 1.2, and they could interest all sectors of the biosphere; for instance, high exergy is a clean lake or river with clean water at low salinity and related inorganic content. On the opposite, low exergy is water contaminated with high salinity and high biochemical oxygen demand (BOD) or chemical oxygen demand (COD). This is applicable also to technological nutrients; high energy means to have copper or lead concentrated in one technological reservoir rather than disperse as waste in the environment, as indicated in Figure 1.6.

Anthropogenic activities bring about a displacement of nutrients from the biosphere to waste.

Nutrients supply plant species with water and resources, stimulating growth rates that in turn increase biomass and chemical exergy yield.

Since the availability of nutrients and water is regarded as the main factor determining growth in the biosphere today, artificial

FIGURE 1.6 Coal life cycle. (Adapted from Cheng, I. et al. 2017. *The Great Water Grab: How the Coal Industry is Deepening the Global Water Crisis.* Greenpeace International. http://www.greenpeace.org/international/en/publications/Campaign-reports/Climate-Reports/The-Great-Water-Grab/.)

displacement of nutrients generated by modern life affects the chemical potential of the same.

In this respect, high exergy is having each element where it must be and where it can contribute to the biosphere grow. As a consequence, not only the ecological value of the soil–water environment affected by the pollution discharge is decreasing but also its exergetic value.

An example of how these could be correlated is indicated in Figure 1.6 whereby the use of coal fuel in power plants is represented throughout all its cycle.

As can be seen from the previous figure, coal combustion not only decreases the exergy value of the Earth as it consumes fossil fuel, a resource stored in the planet, but it also leaves a large impact on the biosphere by contaminating land water and air.

As can be seen from Figure 1.6, the impact of coal plants starts from the excavation for mining or quarrying, which in turns brings about a disruption of natural ecosystems where the extraction takes place, affecting the chemical potential of the local biosphere and its capability of absorbing and storing energy from the sun.

Afterwards, coal ashes resulting from the combustion and escaping from the fumes exhaust treatment are dispersed in the terrestrial and aquatic environment through the exhaust gases.

The flying ashes precipitated within the coal plant sometime are re-employed in road construction and concrete production, but in most cases, they become a toxic waste that is generally stored in large coal ash ponds and coal ash dump sites, practically destroying the ecosystems in the area surrounding the ponds.

In addition, expensive additives are often used in the steam process to avoid hot gas path deposition and exhaust gas emissions, and these potentially expensive technological nutrients (McDonough et al., 2013) dispersed in the atmosphere not only are polluting but are lost for further anthropogenic cyclic use.

Because of the large quantities of water used for quarrying, washing, and cooling the water, the ecosystem cannot support the growth of biomass in an equal manner; therefore, in this part of the book, it is demonstrated that the corresponding exergetic impoverishment of the Earth is not only deriving from the use of coal and its abstraction and consumption but also from impoverishment of the chemical potential of the elements in the biosphere.

Clearly, the considerations made previously are not only related to coal power plant but also involve many of anthropogenic activities such as mining and smelting of metals, burning of fossil fuels, use of fertilizers and pesticides in agriculture, transportation, production of batteries and other metal products in industries, sewage sludge, and municipal waste disposal, which results in a gradual impoverishment of the exergetic level of all ecosystems affected.

In particular, the overall exergetic value of the Earth is diminished as fossil fuel is consumed, and in addition to the exergy dissipated by combustion, the Earth's soil and the air are contaminated and have a lower propensity to accommodate anthropogenic activities.

As a consequence, the biosphere and life in general are depleted and clearly prevented from fully undertaking the chemical redox reaction of photosynthesis that store exergy in the planet and drive all systems downstream.

This obviously decreases or annihilates the chemical exergy load that is applied to the plant–soil system by the solar plant and by the natural cycle of nutrients regeneration and decreases the Earth capacity for assimilation and removal as a whole.

In fact, apart from the abstraction of exergy in form of fossil fuel and biomass, some of the anthropogenic activities through pollution of the air, water, and land decrease the Earth's capability to maintain the living process in the biosphere and, therefore, decrease the potential to generate and store exergy.

Some of the renewable energy sources such as concentrated solar power may also require cooling water; however, the amount of water that is required is limited compared to fossil fuel source, and the impact in terms of pollution of the water stream is substantially lower.

The issue, however, is related to the production of photovoltaic (PV) panels that is taking place in countries where the environmental regulation related to the discharge of residues from silicon to the environment is poorly regulated and often neglected.

As it can be seen from Figure 1.7, land could be viewed as a combined thermodynamic system comprised of two subsystems (natural soil and vegetation). These systems function through energy and mass flow (e.g., solar radiation, water, and nutrients) derived from the surrounding environment.

In such a thermodynamic system, chemical exergy is stored in the soil as organic carbon with microorganisms, and some energy, due to effluent application, undergoes a series of dissipative processes (e.g., decomposition, nitrification, etc.) within the terrestrial environment, releasing various components and heat to the environment. These components are subjected to plant and microorganism uptake, chemical reactions in the soil, or emissions to the atmosphere.

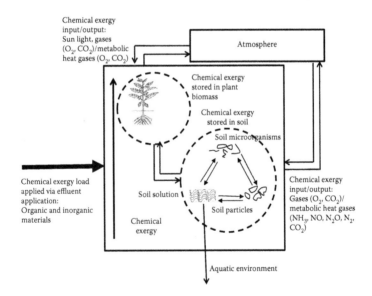

FIGURE 1.7   Schematic illustration of chemical exergy flows throughout the land systems. (Adapted from Tzanakakis, V. A. and A. N. Angelakis. 2011. *Ecological Modelling.* 222: 3082–3091.)

The chemical exergy of the components remaining in the soil, if not provided in excess to the biocapacity of the ecosystem, increases the soil exergetic level.

In the absence of substantial anthropogenic activities, the cycle is providing a gradual increase of the material Earth exergy, generated both by the accumulation of exergy in the biomass and by the increase of the soil exergetic content.

The value of the chemical exergy of these components is based on the concentrations of organic carbon (as COD), ammonium (NH_4^+), nitrates (NO_3^-), phosphorus (as PO_4^{-3}), sodium (Na^+), calcium (Ca^{2+}), and magnesium (Mg^{2+}).

As indicated in Figure 1.7, this cycle of energy accumulation brings about not only a state of sustainability but a state of dynamic gradual accumulation of chemical exergy accumulated in soil plant and related organisms only partially dissipated through respiration and oxidation. This is a process of abundance.

However, as indicated earlier, this value is destroyed or sharply decreased if the ecosystem is affected by the discharge of pollutants that are incompatible with the environment or are discharged in excess to the environment biocapacity; in this situation, the cycle of abundance is broken.

The discharge and dispersion, for instance, of toxic ashes generated by anthropogenic activities in the soil from thermodynamic point of view—and therefore disregarding the adverse aspects on human health—breaks the natural cycle of soil enrichment and decreases the exergetic value of the land itself.

Figure 1.8 shows coal combustion waste or residuals, a byproduct of coal power plants captured from smoke stacks and stored in large retention ponds.

This approach can be considered not only for the discharge of toxic material but also for the discharge of pollutants not compatible with the natural regeneration accumulation cycle such as the dispersion of plastic bottles and nylon bags in the environment, as indicated in Figure 1.9, or hydrocarbons that subtract soil to the process indicated previously of soil enrichment.

FIGURE 1.8 Coal ash contaminated soil. (Retrieved from www. ombwatch.org.)

FIGURE 1.9 Plastic bags in an uncontrolled dump site. (From Lucas Mullikin, 2016.)

TABLE 1.3   Factors Affecting Earth Exergy Depletion

Item	Description	Consequences
1	Abstraction and consumption of fossil fuel, biofuels, and biosphere energy for human life (food, textile, etc.)	Consumption of exergy stored as fossil fuel decreases the exergetic content of the planet. Biofuel and food production could be done sustainably.
2	Pollution of the environment	The pollution of the environment brings about a gradual decrease of the chemical potential of the biosphere and a consequent impoverishment of Earth's exergetic content.
3	Reduction or impoverishment of natural biosphere and ecosystems	Decrease of active area for exergy accumulation and storage.

It is possible to consider that as the energy of the Earth has been in a dynamic state throughout all the existence of the biosphere, the environmental impact has the effect of decreasing the exergy of the Earth.

The energetic decrease occurs because of three main factors, which are indicated in Table 1.3.

1.6 ABUNDANCE OF ENERGY AS A NEW BENCHMARK BEYOND SUSTAINABILITY

Since the natural pattern, in the absence of anthropogenic activities, foresees an accumulation of energy and resources additional to the dissipative process that supports life, the concept of sustainability does not represent the absolute threshold of environment respect.

Sustainability, in fact, indicates the maintenance of the status quo. Its definition has the objective to ensure future generations with long-term ecological balance based on today's state of affairs. This is a static vision whose target is limiting and not ambitious enough.

The Earth is in a continuous dynamic state. Beyond sustainability there is abundance, which means not only maintaining the status quo but reversing the environmental damage and, even more, allowing the accumulation of natural energy and resources for future generations.

Abundance is a well-known term and has been described several times also in relation to sustainability (McDonough et al., 2013).

However, how abundance can be quantified and where the energy and resource thresholds are that identify when the sustainability concept is passed in a direction or another has not been defined.

Accordingly, it has not been quantified how sustainability thresholds can be identified in numerical terms.

The leading point in this chapter is, therefore, to identify the sustainability threshold beyond which humanity can thrive in a regime of abundance and set this regime as a new benchmark for the sustainability goal.

To this purpose, the exergetic flow diagram indicated in Figure 1.1 has been simplified, as shown in Figure 1.10, taking into account only the in-out net exergy flows, the climate dissipative processes, and the biosphere exergy uptake. It becomes evident from Figure 1.10 that for thousands of years the exergy requirements for anthropogenic activities were negligible, the Earth has been in a continuously dynamic state, and the energetic status of the planet was increasing as the time went by.

The analysis of the same energy balance shows also that the process of exergy accumulation on the planet started with the appearance of the biosphere and, therefore, with the appearance of life on the planet.

However, the Earth, and in particular the biosphere, has never been in equilibrium from an exergetic state; without anthropogenic activities, the Earth was increasing its exergy.

Accordingly, three important conclusions can be drawn from the analysis carried out so far:

1. There is an exergy state on the Earth, and this state is dynamic.

2. Exergy on the Earth continued to increase until the appearance of substantial anthropogenic activities.

3. Exergy was accumulated on the Earth when life appeared thanks to the biosphere and the chlorophyll capability of absorbed solar energy.

As can be seen from Figure 1.10, the Earth does not have an energy problem insofar that the energy the Earth receives from the sun is abundant.

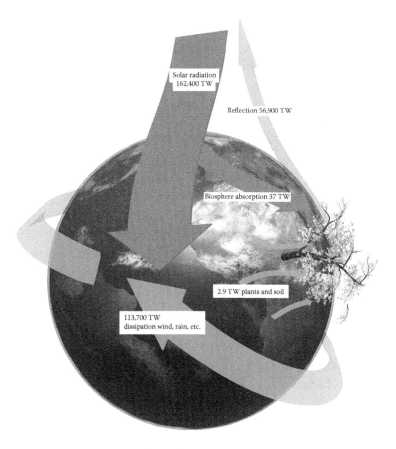

Solar radiation
162,400 TW

Reflection 56,900 TW

Biosphere absorption 37 TW

2.9 TW plants and soil

113,700 TW
dissipation wind, rain, etc.

FIGURE 1.10   Exergy uptake reflection and storage.

Nature has been storing energy in the biosphere for thousands and thousands of years and continues to do so despite impairment by current pollution and reduction of biosphere that has been brought about by humanity.

In particular, the Earth has been in a process of exergy accumulation from the sun and accumulation of chemical compounds in their reduced state starting from their oxidized state. This has been made possible by the photosynthetic process.

In other terms, the Earth was increasing its chemical potential by reducing carbon in the presence of sunlight by a set of redox reactions. The carbon is in turn stored in the biosphere and in the soil.

The process has not been interrupted, but when anthropogenic activities became prominent, the balance between the exergy that is accumulated by the biosphere and the exergy that is withdrawn or destroyed by the anthropogenic activities waste became negative.

As indicated previously, energy and sustainability are traditionally associated with the abstraction and use of fossil fuel and non-renewable energy. In reality, this chapter of the book describes how the current anthropogenic activities on the planet and the environmental impact have a much wider impact on the exergy disruption.

The decrease in energy is taking diverse forms from the obvious decrease in fossil fuel level, to the decrease of the energy stored in the biomass, to the decrease of organic carbon and nitrogen fixed in soil that decrease the soil chemical potential.

The issue, since industrial activities in the Earth became prominent, rather than an energy problem is a negative energy balance. This means that the anthropogenic activities consume and destroy with their pollution more energy than nature is capable of storing.

The result of this is a constant decrease of Earth energy since the beginning of industrial era.

To identify and quantify a threshold for sustainability and a criterion for abundance, it is necessary to make an energy balance between the energy that is withdrawn or destroyed and the one that has been accumulated.

With reference to the schematic diagram indicated in Figure 1.10, the evolution of the exergy state on the Earth throughout time is indicated by Equation 1.2:

$$\dot{E}_S = \dot{E}_r + \dot{E}_{plan} + \dot{E}_{bio} \tag{1.2}$$

where

$\dot{E}_S$ is the exergy flux received by the cosmos through its radiation.

$\dot{E}_r$ is the exergy flux reflected back to the cosmos.

$\dot{E}_{plan}$ is the exergy flux dissipated to support the process of transformation that normally occurs in nature on a normal planet.

$\dot{E}_{bio}$ is the exergy requirement to drive the biological activities in the planet and a share of the term. $\dot{E}_{bio}$ which can be defined as γ is accumulated in chemical energy through photosynthesis, nitrogen fixation, etc.

As the term $\dot{E}_r$ is reflected back, the term $\dot{E}_{plan}$ is dissipated, and what is accumulated in the Earth is the fraction γ of $\dot{E}_{bio}$.

Therefore, before the advent of the industrial era and neglecting for simplicity and reactions of chemical elements that occur outside the biosphere, the exergy variation of the Earth dE_x over the time dt can be indicated by Equation 1.3:

$$\frac{dE_x}{dt} = \gamma \cdot \dot{E}_{bio} \geq 0 \tag{1.3}$$

which can be explained in simple words as follows: the variation of Earth exergy with time was equal to the energy that the biosphere was able to store.

In this case, without the presence of sensible anthropogenic activities, the term $\gamma \cdot \dot{E}_{bio}$ is generally positive; this implies that the biosphere has been accumulating energy and was thriving into a state of abundance. This has happened for thousands and thousands of years until the advent of the industrial age.

With the advent of large industrial activities, energy started being withdrawn as fossil fuel. At the same time, the biosphere started shrinking, and pollution affected the components of the biosphere; therefore, the balance indicated in Equation 1.4 becomes:

$$\frac{dE_x}{dt} = \gamma \cdot \dot{E}_{bio} - (\dot{E}_{extr} + \Delta E) \tag{1.4}$$

where

E_x is the exergy value of the Earth,

$\dot{E}_{extr}$ is the exergy dissipated from the planet exergy reserves (both fossil, chemical, and biological; biosphere; and food) for anthropogenic activities, and

ΔE is the biosphere exergy dissipated by the pollution generated by anthropogenic activities.

This formula can be translated in the following terms: the variation of Earth exergy with time is equal to the energy that the biosphere is able to store minus the exergy withdrawn as fossil fuel and the exergy dissipated by discharge of pollutants in the environment and by the shrinking of the biosphere and its energy.

This time, the exergy variation can be either positive or negative, and it depends on the difference between the exergy stored in the biosphere $\gamma \cdot \dot{E}_{bio}$ and the sum of the exergy withdrawn and destroyed.

It is possible, therefore, to have different scenarios, which are described below:

A scenario of abundance: In this case, the energy from the cosmos is accumulated in the planet in the form of biomass, etc., at a higher rate than these resources are consumed.

In this case, $(dE_x/dt) > 0$: The Earth is storing energy coming from the sun.

A *scenario of non-sustainability*: In this case, through the anthropogenic activities, energy resources are drawn from the Earth and are dissipated at a rate that is higher than the rate these resources are replenished. In this case, factor $(dE_x/dt) < 0$ means that the Earth is drawing from its natural resources that are used as storage of energy (i.e., biomass fossil fuels) to support its living requirements.

The energy sustainability threshold, therefore, is when there is no variation in the exergy level of the Earth with time, and therefore $dE_x/dt = 0$. The biosphere in this scenario accumulates the same energy that humanity dissipates. In this case, we have the sustainability threshold defined in Equation 1.5 as:

$$\gamma \cdot \dot{E}_{bio} = (\dot{E}_{extr} + \Delta E) \tag{1.5}$$

Table 1.4 below summarizes the scenario that have been previously described.

TABLE 1.4   Energy Sustainability Indexes

Item	Term	Situation	Note
1	$\dfrac{dE_x}{dt} = 0$	Equilibrium sustainability threshold	The Earth is thermodynamically sustainable. It receives as much energy from the sun as the energy consumed for its internal life supporting systems.
2	$\dfrac{dE_x}{dt} < 0$	Non-sustainability	The Earth is thermodynamically non-sustainable. It receives less energy from the Sun than the energy consumed for its internal life supporting systems, and therefore it draws from its exergy reserve.
3	$\dfrac{dE_x}{dt} > 0$	Abundance	The Earth in a state of abundance whereby it accumulates exergy.

1.7 DISRUPTIVE METEOROLOGICAL EVENTS AND ENERGY UNBALANCE

As indicated, the net energy received from the sun ($\dot{E}_S - \dot{E}_e$) is either dissipated in meteorological events and planet heat, or it is absorbed by the biosphere.

This concept is illustrated in the energy balance by Equation 1.2:

$$\dot{E}_S = \dot{E}_r + \dot{E}_{plan} + \dot{E}_{bio}.$$

It is not the aim of this publication to make reference to climate changes that have been observed in the last years. On the other hand, as effect of the greenhouse emissions, the net energy flow to the Earth from the sun increases as less energy is reflected $\dot{E}_r$; in addition, a reduction of biosphere energy absorption brings about a corresponding increase of the energy that needs to be dissipated through the other mechanisms, including meteorological events.

In other words, the energy dissipated in meteorological events and planet heat $\dot{E}_{plan}$ equals the net energy received from the sun minus the energy absorbed by the biosphere.

This in turn means that the lower $\dot{E}_{bio}$ the greater $\dot{E}_{plan}$, so the lower the biosphere, the higher the energy that goes into heat absorption but also rain, wind, etc., in addition to the higher net energy flow that is received.

It is impossible to make a rigorous comparison, but it has been observed that the reduction of the biosphere has been concomitant to the increase in the violence of some meteorological events This seems to find a correlation based on the exergy balance.

1.8 ENERGY SUSTAINABILITY: NOT ONLY FREEDOM FROM FOSSIL FUEL

As indicated earlier in the book, a time will come when technology will be able to generate all humanity's needs from renewable energy at a reasonable cost. In this situation, it will be possible to live with what remains of the fossil fuel reserves perhaps simply as a strategic storage (McDonough et al., 2013).

However, the analysis indicated previously shows that in this case humanity may not reach a sustainable energy situation.

The sustainability threshold is defined by Equation 1.6 as:

$$\gamma \cdot \dot{E}_{bio} = (\dot{E}_{extr} + \Delta E) \qquad (1.6)$$

In fact, in the case of $\dot{E}_{extr} = 0$, which means all energy required by humanity is produced by renewable source, the energy challenge still remains of ensuring that the term ΔE, the exergy that humanity dissipates through anthropogenic activities and environment pollution, is lower than or equal to the energy that the biosphere can accumulate.

This therefore leads to a great emphasis to the preservation of the biosphere and its energy content in terms of biomass, soil organic components, and its health through prevention of pollution and waste emissions that destroy the exergy levels of the biosphere.

1.9 RENEWABLE ENERGY: REGENERATE PRISTINE CONDITIONS AND BUILD MORE ABUNDANCE

With the development of renewable energy resources, new potentials for energy that utilize the exergy available from the natural processes occurring in nature (wind, tidal hydro energy projects, etc.) have been made possible.

Therefore, the schematic arrangements of Figure 1.1 (Szargut, J. T. 2003) need to be revisited. Consequently, the scheme in Figure 1.11 has been developed from the original model.

Figure 1.11 considers recent renewable energy developments, indicated in green. As can be seen from Figure 1.11, renewable energy sources such as wind tidal or hydropower take advantage from the exergy destruction of natural phenomena to produce energy for human purposes.

This is an enormous step towards the goal of sustainability, particularly as renewable energy taps into an energy stream that is much more abundant than the energy that goes into the biosphere or that is stored as a fossil fuel.

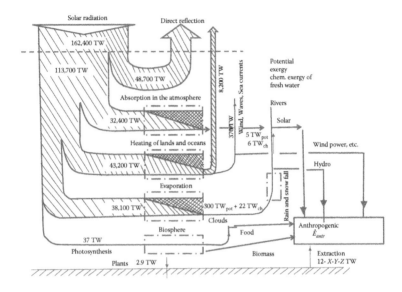

FIGURE 1.11   Renewable energy resources input to anthropogenic activities and overall exergetic balance of the Earth. (From Szargut, J. T. 2003. *Energy.* 28(11): 1047–1054.)

As can be seen in Figure 1.12, the potential for renewable energy is by far greater than the energy requirement for anthropogenic activities.

Renewable energy reduces the energy dissipated by the discharge of pollutants in the environment. Furthermore, solar panels are often installed in deserted areas or building rooftops. In these conditions, renewable energy does not generate any shrinking of the biosphere, which can continue its function of energy absorption through oxidation-reduction (redox) reactions and storage capacity.

If this energy could be harvested to the levels required to sustain anthropogenic activities and reverse environmental pollution, we would have from an energetic point of view a situation $(dE_x/dt) > 0$, which means a positive variation of the exergy of the planet across the time.

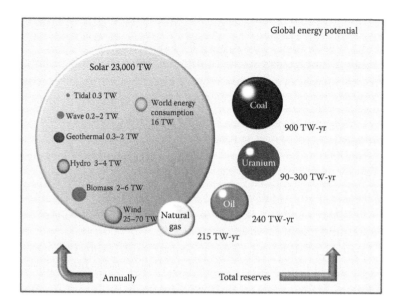

FIGURE 1.12   Global energy potential and reserves. (Adapted from Shahan, Z. 2014. http://www.renewableenergyworld.com/ugc/articles/2014/11/5-reasons-solar-power-will-dominate-energy-in-the-next-century.html.)

This means a situation of abundance where, as the time goes by, the biosphere is left to clean the planet thanks to redox reactions that remove oxidized pollution products from the Earth while anthropogenic energy needs are satisfied by renewable sources.

This is a process that accumulates and stores in the Earth elements with a high biochemical potential that constantly increase the planet's exergy.

Considering the order of magnitude of the energy streams involved, this objective could be easily in mankind's reach regardless of the exergy requirements necessary to sustain the anthropogenic needs.

In principle, as the time goes by, the biosphere could be left to reinstate the exergy level that we had before the advent of industrialization, and humanity could thrive towards an infinite exergy storage that is usable to endless purposes.

Energy storage can occur in different manners, and recently humanity has been able to directly or indirectly accumulate energy. Batteries, gravity storage, hydropower, and in future hydrogen from renewable sources can provide a useful energy storage for immediate anthropogenic use.

However, a significant and essential component of the exergy is effectively stored in the planet in the form of organic energy associated with the chemical status of the elements that compose the biosphere, and this is ensured by the absorption of energy by plants and forestation in the biosphere.

Figure 1.13 shows the Leuser Ecosystem in Aceh, Indonesia. This is one of the densest forests on Earth that segregates millions of tons of carbon every year, generating oxygen and storing over 1.6 billion tons of organic carbon (Shah, 2017).

FIGURE 1.13 The Leuser Ecosystem in Aceh, Indonesia. One of the densest forests on Earth. (Shah, V. 2017 NGO and PepsiCo feud over deforestation, labour claims, http://www.eco-business.com/news/ngo-and-pepsico-feud-over-deforestation-labour-claims/.)

1.10 LOOKING BACK AND LOOKING FORWARD

The evolution of the exergetic state on the planet can be schematically described, as illustrated in Figure 1.14. Exergy was stored and gradually accumulated in the Earth when life appeared with its unique capacity of using cosmic radiation to reduce carbon into organic compounds. Thanks to the biosphere, the exergy levels in the Earth kept accumulating and increasing until the exergetic requirement of the anthropogenic activities became comparable and higher than the natural exergy storage mechanics. This process was further accelerated by the gradual deforestation and disruption of natural elements that compose the biosphere and energy of the plant decreases.

This exergetic growth was accompanied by a constant increase of the oxygen concentration in the biosphere.

The future may present three different scenarios that are different according to how responsively humanity is capable of administering the planet energy and biological resources.

The first scenario maintains a neutral position and corresponds to the horizontal line in Figure 1.14. This could be called a

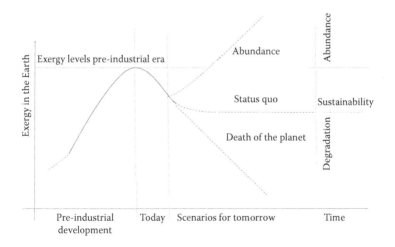

FIGURE 1.14 Development of the Earth exergy level and future scenarios.

sustainability scenario as the exergy of the planet neither increases or decreases.

As seen in Figure 1.14, in the second scenario, the energy accumulated and stored is greater than the energy consumed.

This means in turn a condition whereby the exergy of the planet is constantly increasing thanks to redox reactions that accumulate and store in the Earth elements with a high chemical potential.

This objective could be easily in mankind's reach regardless of the exergy requirements necessary to sustain the anthropogenic needs.

The third scenario would be the opposite. This scenario would occur if the exergetic level in the Earth gradually decreased to a minimum state until no reactions among the elements in the Earth would be possible. This corresponds to the exergetic death of the Earth that would lead to conditions whereby no life would be possible.

There is an association that between biosphere health and planet exergy. The destruction of the exergy of the planet corresponds to a degradation of the environment, and restoration of the environment corresponds to an accumulation of exergy, a resource for future generations.

However, the sustainability threshold can be overcome in both directions, and fortunately, environmental impact is reversible.

In this respect, a judicious use of the energy the Earth constantly receives from the sun can reverse a negative environmental impact and provide exergetic storage for the planet to reserve for future generations.

1.11 CLOSING THE LOOP: CHOICE FOR ABUNDANCE INSTEAD OF EXERGETIC DEATH OF THE EARTH

As indicated in the previous sections of the book, the planet would degrade to a state of energy decay without constant energetic and exergetic flux from the sun. The exergy received from the sun is in turn partially reflected back to the universe, partly absorbed and stored, and partly used to support life on the planet.

In principle, the Earth is open to the exchange of mass with the rest of the universe. Since the Earth receives cosmic rays, particles, and meteorites, while at the same time emitting atom particles and, from time-to-time, satellite and space shuttles, it could be considered an open system.

On the other hand, it is reasonable to consider this exchange of mass negligible; therefore, the Earth can be effectively considered a system where mass is practically constant but where energy can be exchanged.

In fact, all types of energy exchange are allowed, and the Earth receives from space radiation of all types and kinds, and it emits radiation.

Therefore, from a thermodynamic point of view, the Earth interfaces with the rest of the universe and can be, in principle, considered a closed system consisting in turn of two subsystems, as schematically indicated in Figure 1.15.

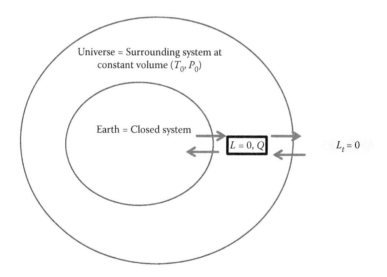

FIGURE 1.15 Thermodynamic scheme of the Earth as a closed system. (Adapted from Chen, G. Q. 2005. Exergy consumption of the Earth. *Ecological Modelling.* 184: 363–380.)

The first subsystem, the Earth, is a closed one capable of exchanging exergy in the form of heat with the second subsystem, the universe, which remains at constant temperature T_0 and pressure P_0.

It is considered in this case that the volume of the surrounding system is constant, and, therefore, there is no exchange of heat through the total system.

There are several theories that elaborate on modern society energy and material dissipation and their inherent exergetic disruption. These were set up in conjunction with the Second Law of Thermodynamics (Capilla et al., 2014).

In this part of the book, the concept of sustainability is directly related to the Second Law of Thermodynamics and to the irreversibility that is associated with every transformation occurring in nature.

As a closed system, the Earth varies its internal exergy level via exchanges of work and heat with the surrounding environment and with the dissipation of the exergy destroyed in the processes occurring in nature.

Combining this concept and the first principle of thermodynamics for a closed system, as described in Figure 1.15, the infinitesimal exergy variation of the Earth's internal energy between two states can be obtained by Equation 1.7:

$$dE_{ex} = \left(1 - \frac{T_0}{T}\right) dQ - T_0 \cdot dS_S - dL + p_0 dv - \sum_i \mu_{i0} \cdot n_i \quad (1.7)$$

In this case, dE_{ex} is the infinitesimal variation between two exergetic states of the Earth, T_0 and P_0 are the temperature and pressure of the system surrounding the Earth, in this case the universe.

In this equation, dS_s is the entropic production of the Earth related to the irreversibility occurring in nature and generated by anthropogenic activities. The rest of the terms are as defined in Equation 1.1.

The term dL in this case is the work exchanged between the closed system Earth and the universe, and it is equal to 0. The term $p_0 dv$ is also equal to 0 as there is no volume variation in the physical Earth that can be significant.

In simple terms, this equation states that the exergy variation of the Earth dE_{ex} is equal to the net energy flux received by the external system (in this case the cosmos), which is indicated as: $(1 - (T_0/T))dQ$ minus the exchange of work dL and the variation of volume of the system $(dL, p_0 dv)$, minus the energy dissipated $T_0 dS_s$ and the variation of the sum of chemical potential of the elements in the Earth system multiplied by number of their relative moles: $\Sigma_i \mu_{i0} \cdot n_i$.

Rearranging the terms, the equation becomes Equation 1.8:

$$dE_{ex} = \left(1 - \frac{T_0}{T}\right) dQ - T_0 \cdot dS_S - \sum_i \mu_{i0} \cdot n_i \qquad (1.8)$$

As the energy that the Earth receives from the cosmos, dQ, is constant, the variation of exergy within two infinitesimal times in this case becomes Equation 1.9:

$$\frac{dE_x}{dt} = \frac{-d(T_0 \cdot dS_S + \sum_i \mu_{i0} \cdot n_i)}{dt} \qquad (1.9)$$

Equation 1.9 can be interpreted as another expression of the abundance theory, indicated previously by Equations 1.3 and 1.4:

$$\frac{dE_x}{dt} = \gamma \cdot \dot{E}_{bio} - (\dot{E}_{extr} + \Delta E) \qquad (1.10)$$

where the term $-(\dot{E}_{extr} + \Delta E)$ equals the term $-d(T_0 \cdot dS_S)/dt$ and indicates the dissipation of exergy that occurs on the planet as a consequence initially of the natural phenomena such as wind, waves, etc.

As time went by, and since industrialization commenced, this term includes also the energy dissipation related to the anthropogenic activities.

Therefore, $\gamma \cdot E_{bio}$ equals the variation of the chemical potential of the element composing the planet and demonstrates that life, through the biosphere, increases the exergy of the planet.

1.11.1 Life Upcycles

Abundance versus exergetic death is a choice that humanity has. It is a choice between harvesting and increasing the exergy of the Earth in a process of evolution that can be achieved, whereby each day the planet could potentially store more energy and become better, stronger, and more stable.

The Second Law of Thermodynamics states that as energy dissipates, systems are degrading towards an increased state of disorganization and ultimately to chaos and entropy. Only life, and therefore only the biosphere, prevents this entropic degradation from extending to our planet, by capturing sunlight and developing resources into a higher level of order and complexity. This state of organization and succession, the opposite of entropy, is called negentropy (Schrödinger et al., 1945): a system that transforms the non-living to living, the simple to the complex, the inefficient to the efficient, and represents a process of evolution of the planet.

This is the context in which planet evolution may occur through a continuous accumulation of exergy though various forms of life.

From the appearance of life on the planet, exergy on the Earth continued to grow and kept growing even after the appearance of mankind in the planet.

If life disappeared on Earth, all chemical elements that are a part of the Earth surface, the oceans, and the atmosphere would react among them until no further reaction would be possible, and then the status of zero exergy would be achieved. In this situation, the planet would become too arid and unsuitable for any further life; this would equal to the energetic death of the planet.

CHAPTER 2

Resources and Sustainability

2.1 A NATURAL CYCLE OF RESOURCE ABUNDANCE

Abundance of energy represents a potential to generate abundance all ways downstream in the planet; therefore, it can drive abundance of water, abundance of nutrients (organic and technological), abundance of oxygen, abundance in nature accessibility and health and in humanity's quality of life. All this is possible if energy is managed judiciously.

The leading idea of this chapter, though, is to correlate abundance of energy to abundance of resources and illustrate how this occurs in the natural cycle and how this is affected by anthropogenic activities.

If there were no anthropogenic activities that introduce elements that are alien to the biosphere, it would be possible, in principle, to consider our planet as one immense biological and chemical reactor where the biological and chemical elements of the biosphere are constant but in continuous transformation from one form to another, as schematically indicated in Figure 2.1.

One of these elements is, for instance, oxygen, which is continuously changing from reduced form (O_2) to various oxidized

FIGURE 2.1   Simulation of the Earth as a continuous stirred reactor.

forms (CO_2) in combination with other elements of the biosphere, like carbon.

As it will be demonstrated in this chapter, the Earth is in constant evolution, and in the absence of anthropogenic activities, it would evolve to a state of a gradual accumulation of energy in the form of substances at higher chemical potential.

In this situation, the concentration of one element in the Earth would be regulated by the mass balance Equation 2.1:

$$V \cdot \frac{d\omega}{dt} = \pm \mu \cdot \omega V_1 \qquad (2.1)$$

The equation basically indicates that the variation in the concentration of a component ω, in the Earth (of a volume V), during the interval of time dt is equal to plus or minus the velocity this component is generated or consumed by the biosphere V_1.

In consideration of what has been stated previously, if the natural velocity of waste absorbing and regeneration μ_{rem} (s^{-1}) and a natural velocity of waste emission μ_{em} (s^{-1}) are considered, the term μ indicated in the earlier equation is the net difference between the generation and the absorption of waste (Equation 2.2):

$$\mu = \mu_{rem} - \mu_{em} \tag{2.2}$$

The leading point of this part of this book is that, in the absence of sensible anthropogenic activities, the concentration of a pollutant in the atmosphere would be constantly decreasing, and the concentration of a resource at high redox, therefore at high chemical potential, would increase.

In other terms, as the average rate of pollutant absorption (μ_{rem}) is always higher than the rate of pollutants generation μ_{em}, the term μ is always positive when the component ω is a resource, and it is always negative when the component ω is a pollutant.

Without anthropogenic activities, the Earth would be constantly in a stage of gradual evolution.

Figure 2.2 shows that, in fact, the formation of biosphere on Earth played a very large role in the buildup of oxygen (a chemical element in reduced form therefore at high chemical potential) in the environment. As early as 3.5 billion years ago, some early cells provided some initial form of photosynthesis and produced oxygen as a waste product of their activity.

Initially, the oxygen produced by photosynthetic prokaryotes would have been consumed in chemical reactions or remained dissolved in the oceans, but with additional photosynthesis beyond this level, oxygen would likely have accumulated in the atmosphere.

Producing oxygen in reduced form means in turn storing chemical exergy. This was only possible thanks to the biosphere and the appearance of life on the planet. Obviously, there is a correlation between the increase in the free liberated O_2 and the burial of organic carbon. The burial of organic carbon increases

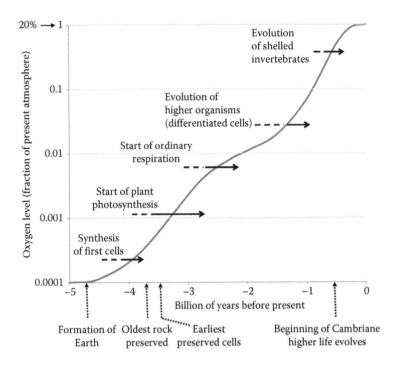

FIGURE 2.2 Oxygen level concentration in the atmosphere (From History of Life on Earth: http://slideplayer.com/slide/2861079/.)

atmospheric oxygen, as the liberated O_2 cannot recombine with the carbon from which it was originally split.

Correspondingly, if CO_2 is considered a pollutant, its concentration on the planet should have decreased along with the increase of oxygen concentration since the appearance of life on the planet.

This is only partially true. The atmospheric CO_2, in fact, has not been always decreasing as the result of biosphere uptake but has increased from 180 ppm during glacial (cold) up to 300 ppm during interglacial (warm) periods.

In this case, while the variations in atmospheric CO_2 from glacial to interglacial periods were caused by decreased ocean carbon storage, these were greatly compensated by increased land carbon storage (300 to 1,000 PgC) and, therefore, by the biosphere action.

There is not a correspondingly long historical trend for organic carbon stored in either biomass or soil, but Figure 2.3 shows the dynamics of carbon stores in a northeastern spruce-fir forest after an initial clear-cut (Ingerson, 2007).

An undisturbed forest continues to build new carbon stores well past a stand age of 125 years. Even though the rate of carbon sequestration may be faster in younger stands (the slope of the total carbon curve is steepest between 25 and 35 years post clear-cut), older forests do continue to add substantial carbon stores each year, as the total carbon line is still rising rapidly at 125 years (Ingerson, 2007). The relevant carbon pool in this case is not only the biomass stored in the tree but includes also dead wood, litter, soil and wood products, and it keeps storing carbon despite respiration and oxidation of compounds before some of it is buried.

Photosynthesis produces buried organic matter that cannot recombine with carbon to form CO_2, and thus must remain in the atmosphere.

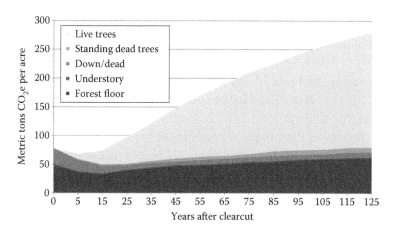

FIGURE 2.3 Non-soil forest carbon, northeast spruce-fir stand. (From Ingerson, A. L. Global Footprint Network. 2007. *U.S. Forest Carbon and Climate Change.* The Wilderness Society: Washington, DC.)

Since carbon dioxide is responsible for global warming, planting trees is the only reasonable and natural way of decreasing this carbon concentration. Humans and animals breathe in oxygen and breathe out carbon dioxide. Trees absorb carbon dioxide and breathe out oxygen. The synergy between humanity and forestation is a natural symbiosis whereby carbon molecules are passed on each time in a manner that makes it easily reclaimed by the breathing of carbon in the atmosphere, biosphere, and so on until they get back to soil or atmosphere where it gets stored and upcycled by the photosynthetic process.

In this situation, carbon becomes part of the soil in the end and is not contaminated by materials that toxify living systems (McDonough et al., 2013).

In addition, trees allow the ecosystem to flourish and allows the biosphere to enrich, enabling the storage of more biochemical energy, as indicated in Equation 1.1.

Planting trees not only makes sense from the climate change point of view but more generally for an overall new approach to life and nature as they provide a home to wildlife and beautify the environment humanity lives in.

Trees contribute to store solar energy since trees naturally pull carbon dioxide from the air and sequester it for a long time in their cellulose and other structures. In addition, trees generate a gradual increase of the material Earth exergy both by the accumulation of exergy in the biomass and an increase of the soil exergetic content.

Thanks to the unique capabilities of forestry to convert CO_2 to reduced form of carbon through photosynthesis, from the energetic point of view, forests are the exergy generators and storage systems of the planet.

Trees and greeneries, therefore, harvest and store exergy from the sun and distribute this energy in the whole lifecycle downstream.

Figure 2.4 shows a healthy soil rich in organic matter that has been the results of years of organic carbon accumulation.

FIGURE 2.4 Rich organic layers beneath the surface tundra soil. (Adapted from Xanthe Walker/Northern Arizona University.)

The soil is made up of heterogeneous mixtures of both simple and complex substances containing carbon. The sources for organic matter are crop residues, animal and green manures, compost, and other organic materials. A decline in organic matter caused by the reduced presence of decaying organisms or an increased rate of decay because of changes in natural or anthropogenic factors indicates a loss of energy in the soil and a corresponding loss of resources and chemical potential of the soil.

As it can be concluded from the previous equation, since the terms $\pm \mu \cdot \omega \cdot V_1$ is generally always different from zero, the Earth is in a constant and dynamic change, and the concentration of any element in different chemical state is in constant development.

As it could be seen in relation to the previous chapter, since the term dE_x/dt has been greater than zero until anthropogenic activities developed sharply in the last two centuries, then then also the term $d\omega/dt$ indicates the potential for a substantial enrichment of the planet thanks to various reduction reactions which in turn bring about an increase of the chemical potential of the planet. Therefore, in principle, the concept of abundance as it can be seen from Figure 2.1 is well justified from both an energy and a biochemical resources point of view. Thanks to the positive exergy balance from the sun, there is sufficient energy to drive chemical reactions whereby components such as CO_2 can be gradually segregated by the atmosphere and biochemically reduced in chemical compounds that have a higher energy that can be stored as biomass or in the soil.

This is the way to abundance and constant positive dynamic change in the planet.

2.2 NATURAL ABUNDANCE AND ANTHROPOGENIC UNBALANCE

As humanity entered the industrial era, energy and resources (fossil fuels, metals, etc.) from the lithosphere have been abstracted, used, and, as can be seen from Figure 2.5, utilized to supply the anthropogenic requirements. More exergy and resources are abstracted from the biosphere, and these are related to the food, etc.

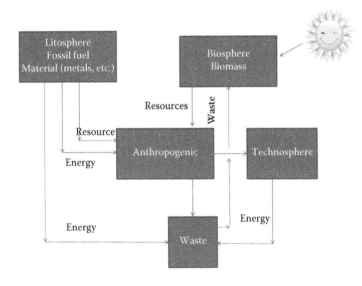

FIGURE 2.5   Flow of exergy and resources in today's practice.

However, not all energy that is abstracted is dissipated to support modern life. A great amount of energy is used to produce goods and infrastructure (buildings, airports dams, power plants, etc.) Therefore, a part of it remains accumulated in those elements that belong to the technosphere or is destroyed in mass consumer goods waste disposal. Metals and mining sectors, for example, spend about one-third of their operating budget on energy that can be recovered by recovering the metals once its technological life is over, and it is disposed.

The biosphere tries to keep up with the waste disposed and uses solar energy, upcycling the organic component, but its regeneration capacity is insufficient compared to volumes that are discharged today. Furthermore, the biosphere upcycling mechanism is very often incompatible with the nature of some of the waste discharged.

Considering both the volume and the concentration of waste that is generated by anthropogenic life, this energy, either in form of chemical energy or heat, cannot be directly disposed in the biosphere and, therefore, requires other energy to be dissipated.

This is one of the great paradoxes in today's waste management practices, either water or solid. Additional energy is required to destroy the energy that remains in humanity's waste, either organic or inorganic.

A typical example is represented by the municipal or industrial waste. Typically, the inherent energy associated with this waste is 15 kJ/g COD (Ioannis Schizas et al., 2004) which would be equivalent to 2.1 kwh/m^3 of inherent thermal energy.

Typically, a conventional treatment plant requires 0.5 to 1 kwh/m^3 to destroy this energy, dissipating the residual waste energy in oxidative processes.

The end of this process is the return of the waste to the biosphere where, even if treated to the standards that have been set forth by the regulatory bodies, the disposed waste again destroys other energy as it reacts with the natural ecosystem and decreases its chemical potential.

As a consequence, more waste is accumulated in the biosphere, whose energy decreases, and more resources are displaced from their original source, dispersed in a form that is not usable or is even toxic for the biosphere. This obviously generates an impoverishment of the biosphere and a decrease in the Earth exergy.

The ecological footprint calculates the combined demand for ecological resources wherever they are located and presents them as the global average area needed to support a specific human activity. This quantity is expressed in units of global hectares, defined as hectares of a bioproductive area with world average bioproductivity (Ecological Footprint Atlas, 2010, p. 13/14).

In practice, the ecological footprint indicates how fast the resources in the planet are consumed compared to how fast nature can absorb the waste and regenerate new resources.

The ecological footprint, in its most basic form, is calculated by the following Equation 2.3:

$$EF = \frac{D_{annual}}{Y_{annual}} \qquad (2.3)$$

where D is the annual demand of a product, and Y is the annual yield of the same product. Yield is expressed in global hectares.

Therefore, the equation of the ecological footprint can also be expressed by Equation 2.4, as indicated in Ecological Footprint Atlas, 2010, page 14:

$$EF = \frac{P}{Y_N} \cdot YF \cdot EQF \qquad (2.4)$$

where P is the amount of a product harvested or waste emitted (equal to D_{annual} above), YN is the national average yield for P, and YF and EQF are the yield factor and equivalence factor, respectively, for the country and land use type in question.

Opposite to the ecological footprint, biocapacity is defined as the total amount of bioproductive land available. "Bioproductive" refers to land and water that supports significant photosynthetic activity and accumulation of biomass, ignoring barren areas of low, dispersed productivity.

In short, it measures the ability of available terrestrial and aquatic areas to provide ecological services. A country's biocapacity for any land use type is calculated by Equation 2.5:

$$BC = A \cdot YF \cdot EQF \qquad (2.5)$$

where BC is the biocapacity, A is the area available for a given land use type, and YF and EQF are the yield factor and equivalence factor, respectively, for the country land use type in question

The chart in Figure 2.6 shows comparatively the estimation of the ecological footprint trends against the biocapacity between

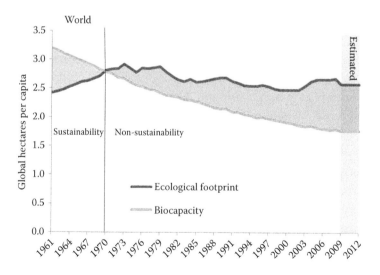

FIGURE 2.6   Evolution of ecological footprint per capita and biocapacity against time. (Adapted from Global Footprint Network, 2013. *The National Footprint Accounts, 2012 edition.* Global Footprint Network: Oakland, CA, page 5.)

1960 to 2010 and the thresholds where, according to this estimation, abundance turned into non-sustainability.

Figure 2.6 shows that the concept of abundance sustainability and non-sustainability indicated in the previous parts of this work can be described in the analysis of the trends of both the ecological footprint and the biocapacity.

For a certain period of time, industrialization could take place without being unsustainable, since the ecological footprint of the planet was lower than the biocapacity. However, Figure 2.6 shows that starting from approximately 1970, the trend was reversed, and the contemporary effect of an increased ecological emission and decreased biocapacity generated a situation whereby the ecological footprint is higher than the biocapacity of the planet.

Until 1970, on average, the biocapacity was higher than the ecological footprint of the planet; therefore, despite the increase in anthropogenic activities, the Earth was in a situation of abundance whereby the energy that was collected in terms of biological products was higher than the amount of pollutants generated. As per the trend indicated in Figure 2.6, the ratio between the ecological footprint and the biocapacity has become non-sustainable starting from 1970 after both a gradual deforestation (a decrease of biocapacity) and a contemporary increase of the environmental pollution (an increase in ecological footprint).

2.3 WHAT NATURE CAN UPCYCLE AND NOT

Referring to Figure 2.1, the situation of equilibrium occurs when the rate of formation of one pollutant equals the rate of removal of the same, and, therefore, there is no accumulation in the reactor environment.

As the Earth does not have any element entering or exiting the reactor, the formula becomes the concentration of the pollutant S and of the biological element ω that is responsible for removing that pollutant governed by the following set of equations, which can been applied as a typical model equation for a continuous stirred tank reactor:

$$V \cdot \frac{d\omega}{dt} = +r_{mg} V_1 \qquad (2.6)$$

$$V \cdot \frac{dS}{dt} = -r_{sr} V_1 \qquad (2.7)$$

where r_{mg} is the velocity of growth of the biological element ω, and r_{sr} is the velocity of the biological nutrient S removal. In this case, the biological nutrient S is also the biological pollutant resulting from biosphere living activities. V_1 is the volume of the biosphere, and V is the Earth volume.

This equation can be explained in simple terms: the variation of the concentration of the pollutant S over the time is equal to the velocity of removal of the same.

As can be seen from the previous equation, biomass ω grows at the expense of the nutrients (pollutants) S present in the feed stream. The basic theories for biological growth are usually based on the Monod Equation 2.8 (Monod et al., 1949):

$$\mu = \frac{\mu_{max} \cdot S}{K_S + S} \qquad (2.8)$$

where μ is the microorganism specific growth rate (s^{-1}), μ_{max} represents the maximum specific biological growth rate (s^{-1}), K is the half-velocity constant (kg/m^3), and S is the limiting substrate concentration (kg/m^3). The μ_{max} value depends on the type of microorganisms as well as on the environmental conditions (e.g., operating temperature).

The rate of microorganism growth (mg/m^3 s^{-1}) is defined in turn as:

$$r_{mg} = \mu \cdot \omega \qquad (2.9)$$

where ω is the biomass concentration (mg/l) digesting or metabolizing the nutrient S (mg/l).

It is possible to assume that the ecosystem behaves as a fluid of volume V_1 (m³), and that the fluid is sufficiently well mixed, and V represents the volume of the bioreactor (m³); S and ω the mass concentration of the element (kgel/kg). The term μ indicates the natural velocity of reaction of the element in the system (s⁻¹).

In this case, there are two types of environment reaction to the discharge of pollutants in the ecosystem.

The first one occurs if the pollutant discharged is compatible with the biosphere, and the biological pollutant removal speed is equivalent to the pollutant discharge.

The pollutant, therefore, can be recycled in the ecosystem, naturally becoming a nutrient for various other biological cycles that, taking advantage of the solar energy, reinstate the original equilibrium and the pollutants concentrations in nature.

In this case, the term is:

$$V \cdot \frac{dS}{dt} = 0 \qquad (2.10)$$

However, we may have a situation whereby the discharge is so high that the natural process of pollutant degradation is too slow, and the biosphere and ecosystems cannot cope with the flow rates discharged.

This is the case of instance of eutrophication of rivers and seas resulting in algae blooming or red tide effects. In this case, we have a gradual increase of the pollutant discharge and, as a result, a corresponding increase of the biomass that grows trying to handle all the nutrients discharged.

We have this as a case where:

$$V \cdot \frac{dS}{dt} \geq 0 \qquad (2.11)$$

and

$$V \cdot \frac{d\omega}{dt} \geq 0 \qquad (2.12)$$

In water systems, for instance, this situation can manifest as an abrupt development of algae blooming.

An environmental alteration that follows the same principles is the example of CO_2 emissions in the atmosphere. Carbon dioxide can be seen as a pollutant of concentration S, which would normally be recycled (and stored) as a natural nutrient for the biosphere in biomass, soil organic carbon, etc.

Carbon dioxide is removed from the atmosphere thanks to a series of oxidation-reduction (redox) chemical reactions driven by the photosynthetic process, but if the pollutant discharge (in this case CO_2) is in excess to the reactor volume (forestation in the biosphere), a gradual increase of CO_2 occurs.

Carbon has always been exchanged among the Earth's crust, the ocean soil, and the biosphere. All these systems consume carbon dioxide and return carbon molecules in reduced chemical forms. In this respect, carbon dioxide to the solar synthetic process becomes organic.

However, there is the possibility of a third case, whereby what it is introduced in the ecosystem is simply not capable to undertake the chemical reactions that regenerate the pollutant itself, or worse, it reacts with other reagents to the point of becoming even more toxic.

In this case, the term $r_{mg} = \mu \cdot \omega$ is equal to 0 and so also the term $r_{sv} \cdot V$.

In this case, the pollutant stays forever in the cycle, and unless it is extracted, it accumulates in an endless way. At the same time, leachates from this pollutant negatively affect the biosphere.

This is the case that occurs with many non-degradable pollutants such as plastic but many other components such as those indicated in Table 2.1.

Alteration and deterioration in the environment have been triggered by a pollutant discharge higher than the natural breathing capacity of the planet or by the introduction of alien elements in the environment that are not upcycled and cannot be regenerated through the natural process.

TABLE 2.1   Non-degradable Pollutants

Non-degradable Pollutants			
Organophosphorus Pesticides	Organochlorine Pesticides	Polychlorinated Biphenyls	Polynuclear Aromatic Hydrocarbons (PAH)
Dimethoate	Aldrin	PCB's	Benzo(a)pyrene
Malathion	Dieldrin	Aroclor	Naphthalene
	DDT	Tetrachlorobiphenyl	
	Chlordane	Trichlorobiphenyl	
	Eldrin		

This is also the example of plastic concentration in the oceans. Plastic cannot be metabolized by the environment, and despite the fact that plastic may be abraded and through erosion transformed into smaller particles, it does not decompose into natural constituents through an ecological process.

Plastic concentration in the environment, therefore, simply increases as the time goes by as more plastic is dumped in the ecosystem. In addition, plastic leaches toxins in the biosphere, which are ending up in food and other nutrients for the biosphere.

After many years, erosion of the marine movements abrades the plastic into smaller components that are ingested by marine animals and transferred to the tissues of the animal but do not change their chemical form.

An ecosystem can be considered as an open system subject to discharge of pollutants from various sources from the environments surrounding it and, of course, from the anthropogenic activities.

However, when anthropogenic activities are introduced, it is important to differentiate in the type of waste and therefore the related pollution. It is possible to have excessive "biowaste." This means the presence of waste that generates pollution in normal concentrations would be perfectly handled by the biosphere, which would at the same time recycle the nutrients contained in this waste.

In this case, there are two harmful effects: the pollution that is generated in the environment and the removal of vital nutrients and species from the biosphere, which degrades them and makes them unusable to humanity.

However, not all pollutants are biological waste. The most critical form of pollution is the one generated by waste of substances and elements that are alien to the ecosystems and cannot be removed by an ecological cycle.

In this case, only an anthropogenic intervention can reinstate the natural equilibrium.

Based on the analysis indicated previously, Table 2.2 shows schematically the different types of chemical pollutant, their impact, and a few examples.

TABLE 2.2   Classification of Pollution

	Type of pollution	Example
Naturally reversible	Caused by natural substances that are normally regenerated naturally in the biosphere, discharged in excess to the ecosystem breathing and regeneration capacity	Carbon dioxide (CO_2) generated anthropogenic activity Nitrogen phosphate in waste water
Naturally irreversible	Caused by natural substances that the biosphere is not capable of regenerating and upcycling when dispersed in the environment	Heavy metals: Lead, mercury
Naturally irreversible	Caused by artificial substances that are not only incompatible with the natural ecosystem regeneration cycle but that can become toxic or leach toxins in the environment	Plastic Pesticides Herbicides derived from trinitrotoluene may have the impurity dioxin, which is highly toxic Synthetic insecticides, such as DDT

The classical definition of a pollutant is the one that describes it as a substance or an energy present in a concentration greater than in nature as a result of human activity that causes detrimental effects to the environment.

However, this definition describes the naturally reversible pollutants such as CO_2 but would not be valid for those substances that do not exist in nature and that nature cannot upcycle.

Chemical elements can link in more complex forms that can be harmful to the environment and to life and therefore become pollutants. To define a strategy for sustainability, it is important to distinguish between each type of pollutant to define a removal strategy.

To identifying the methods and the strategies for pollutants removal or upcycling, Table 2.2 shows a schematic classification of the pollution elements based on their presence in nature and on natural capabilities of upcycling through the ecosystem bioprocesses.

The definition of the sustainability approach therefore must be designed against the different types of pollutants indicated in the previous table and separately analyzed.

2.4 ABUNDANCE OF RESOURCES AS A NEW BENCHMARK BEYOND SUSTAINABILITY

2.4.1 Naturally Reversible Pollutants: A Resource Confused as Waste

In a natural environment, it is not possible to define the concept of pollution because there is no concept of waste (McDonough et al., 2013). The natural ecosystem recycles all matters completely in order to maintain the required level of nutrient. It is also difficult in a natural environment to define the "natural levels" as nature is dynamic; it stores resources and changes as time goes by.

The classical definition of environmental impact occurs when the discharge of one or more components in an ecosystem causes an alternation of the concentration of the same from the natural levels.

Accordingly, components such as CO_2 that are naturally upcycled in nature are considered pollutants in today's perception because their rate of emission exceeds the natural rate of removal.

This is the case of carbon emissions related to climate change but also of many other cases such as waste water polluted with elements that as carbon would be upcycled as nutrients by nature.

In the previous chapter, the natural velocity of waste absorption and regeneration μ_{rem} (s^{-1}) and a natural velocity of waste emission μ_{em} (s^{-1}) were defined. As anthropogenic activities are also going to be considered, it is necessary to define in our analysis another variable, which is the rate of anthropogenic emission of the component considered pollutant.

This term is defined as μ_{ant}, which indicates net specific speed of pollutant generation caused by anthropogenic activities.

Therefore, the overall balance can be defined as indicated in Equation 2.13:

$$V \cdot \frac{d\omega}{dt} = \pm \mu \cdot \omega V_1 \qquad (2.13)$$

where the term $\pm \mu \cdot \omega V$ is always negative in case of a pollutant and becomes:

$$V \cdot \frac{d\omega}{dt} = -\mu \cdot \omega V_1 + \mu_{ant} \cdot V_{ant} \cdot \lambda \qquad (2.14)$$

where V_{ant} is the volume occupied by anthropogenic activities, and λ is an equivalent concentration factor that indicates how intense the concentration of pollutant activities is in the volume of the Earth occupied by anthropogenic activities.

This equation states that the variation of the concentration of a naturally reversible pollutant over a unit of time is the difference between the amount of pollutant we introduce on Earth $-\mu_{ant} \cdot V_{ant} \cdot \lambda$ and the Earth breathing or upcycling capacity $-\mu \cdot \omega \cdot V_1$

In this case, it is important to note that the Earth's breathing capacity is proportional to the volume of the biosphere V_1, therefore decreases as the volume of anthropogenic activities increases. This is equivalent to say, in mathematical terms, that the rate of pollutant removal decreases if the biosphere decreases.

As indicated previously, this process of regeneration of pollutants and upcycle of waste as a resource is possible to a certain extent thanks to many endothermic reactions that occur on Earth and are sustained by the cosmic energy that the planet regularly receives. This is the environmental tolerance to natural waste.

Therefore, it is possible to state that for naturally reversible pollutants, the concept of sustainability occurs when the waste generated by anthropogenic activities reaches an order of magnitude whereby the natural regeneration cycle of the planet is sufficient to eliminate pollutants generated by the biosphere and limited human developments.

Therefore, for this genre of pollutants, it is possible to define a sustainable environmental impact as the one where the level of pollutants introduced in the environment is removed at the same rate as the pollutant is introduced.

In the same way as the exergetic sustainability was analyzed in the previous chapter, the analysis of sustainability for naturally reversible pollutants shows the following scenarios that may occur in practice.

Through anthropogenic activities, the term $V \cdot d\omega/dt$ is positive. This means that the Earth is accumulating pollutants at a rate that is superior to its breathing capacity. In this case, the accumulation of the pollutant is positive, and we have a phase of non-sustainable development.

When the term $V \cdot d\omega/dt$ is negative, pollutants are removed at a higher rate than they are produced, and biochemical reactions that generate components with higher chemical potential occur. In this phase, the Earth is in a process that is beyond the phase of sustainability, whereby resources and energy are stored that can be used for future situations creating abundance.

There is a further condition that occurs when the term $V \cdot d\omega/dt$ equals zero. In this case, pollutants are removed at the same rate as they are introduced in the atmosphere.

This can be defined as the sustainability threshold, which occurs as indicated in Equation 2.15:

$$\mu \cdot \omega \cdot V_1 = \mu_{ant} \cdot V_{ant} \cdot \lambda \qquad (2.15)$$

The three different scenarios are schematically summarized in Table 2.3.

TABLE 2.3 Pollutant Resources Sustainability Indexes

Item	Term	Situation	Note
1	$V\dfrac{d\omega}{dt}=0$	Equilibrium	The velocity at which the pollutant is removed in the environment is equal to the velocity at which the pollutant is introduced in the environment.
2	$V\dfrac{d\omega}{dt}>0$	Non-sustainability	The velocity at which the pollutant is removed in the environment is lower than the velocity at which the pollutant is introduced in the same. A gradual accumulation of the pollutant in the environment takes place and a consequent alteration of the natural cycles.
3	$V\dfrac{d\omega}{dt}<0$	Abundance	The velocity at which the pollutant is removed from the environment is higher than the velocity at which the pollutant is introduced in the environment. The pollutant is converted into a chemical form that is beneficial to the environment and can be used in a subsequent stage through a natural process of regeneration.

In the 1970s, a theory was developed whereby the intensity of the environmental impact should be considered proportional to the technology T, affluence A, and population P levels of a society.

Accordingly, the approach to the pollutants balance in the atmosphere, which was indicated with the previous formula, could be modified as indicated in Equation 2.16:

$$V \cdot \frac{d\omega}{dt} = -\mu \cdot \omega \cdot V_1 + K \cdot P \cdot A \cdot T \qquad (2.16)$$

In this case, the sustainability threshold, whereby the amount of pollution is equivalent to the amount of regeneration, can be identified by Equation 2.17:

$$\frac{K \cdot P \cdot A \cdot T}{\mu \cdot \omega \cdot V_1} = 1 \qquad (2.17)$$

The previous equation states that the population threshold and the level of affluence and technology (A, P, T) of the planet for those components that are naturally upcycled is dictated by the regeneration capacity of the planet $\mu \cdot \omega \cdot V_1$.

Therefore, as long as population, affluence, and technology are consistent to the upcycling rate of the pollutants generated by humanity, a condition of sustainability occurs.

Unfortunately, the parameters in the terms $K \cdot P \cdot A \cdot T$ cannot be controlled, and rightly so. It is not acceptable to dictate the level of population of affluence or technology that humanity has the right to have; therefore, as time goes by, this term of the equation will unavoidably become greater.

On the other hand, the only element that can counterbalance the increase on the $P \cdot A \cdot T$ factor is the volume of the biosphere V_1. Also in this case, with an increasing population, the biosphere is unavoidably shrinking; therefore, the denominator of this equation decreases.

This has been the main contradiction of the previous waves of economic development that seem to suggest that humanity cannot progress unless destroying the Earth. The leading point of this part of the book is that it does not have to be this way, and as technology progresses, the goal of sustainability and abundance can be achieved thanks to technology.

This will be dealt with in the following chapters as technology and the reinstatement of a new abundance and environmental prosperity will be discussed.

2.4.2 Naturally Non-reversible Pollutants: The Subtle Threat

The sustainability threshold for natural waste is constrained by the volume of waste and the biocapacity required to upcycle this waste. However, there are other waste typologies that are generated by anthropogenic activities that cannot be upcycled or are even toxic. For these non-reversible waste, the planet biocapacity is practically non-existent.

As mentioned earlier, sooner or later technologies will be developed capable of harvesting as much clean energy as is required for humanity from renewable sources at affordable cost, reducing the amount of naturally reversible pollution elements.

However, also in this scenario, the challenge of sustainability will still have problems. These are related to non-naturally reversible pollutants.

With reference to Equations 2.6 and 2.7, it should be noted that for naturally irreversible pollutants, the term $V \cdot d\omega/dt$ is always greater than 0, and from the natural ecosystem point of view, this is a situation of constant non-sustainability.

It is well known that with the current level of industrialization and consumerism, planet resources are utilized faster with respect to nature's capability to absorb the waste and regenerate new resources.

This aspect is further aggravated by the gradual reduction of the biosphere volume from deforestation and increased land use for anthropogenic purposes, as schematically indicated in Figure 2.7.

In this case, there will be certainly a phase in humanity's evolution where products that are entirely biodegradable will be developed and commercialized for unrestricted human use, ranging from the food industry to pharmaceuticals or transportation.

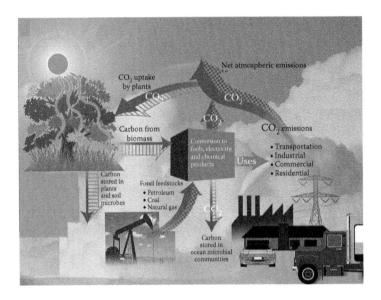

FIGURE 2.7 Biosphere cycles and anthropogenic impact.

One is the concept of dematerialization. This concept is generally developed by industries to increase their competitiveness, and it involves the use of less material per unit of output. This also brings about a consequent reduction in waste generation.

An interesting comparison is indicated in Figure 2.8, which shows the approximate specific weight of construction for the main state of the art power generation technologies.

The data for the graph illustrated in Figure 2.8 are taken from typical plants installed in the Middle East.

As can be seen, technologies such as coal and oil-fired power generation burn in one year a fuel mass more than 20 times larger than their construction weight, and they, therefore, generate an equivalent amount of waste.

The same mass must be piped or transported, sometime from hundreds (in the case of coal thousands) of kilometers from the point of use.

At the end of their life, some of the plant construction mass is recycled as scrap, but an enormous amount is wasted in corrosion

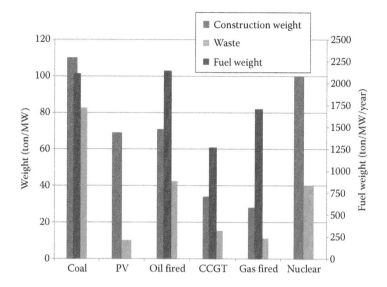

FIGURE 2.8 Approximate construction and yearly operation mass required and wasted in different power generation technologies.

and erosion products or is heavily oil contaminated and then unrecyclable.

Photovoltaic (PV) technology currently has a quite high construction mass (75–80 tons/MW installed for GCC countries), but it does not require any significant further mass for operation.

Most (about 85%) of the PV plant mass is recycled at the end of the lifecycle. Furthermore, it should be considered that the previous data do not include the weight of civil works. Also in this case, concrete requirements are substantially higher for coal and oil fire technologies than for PV.

Lower mass means lower energy for construction and lower waste. This does not undermine in any respect the grave problems encountered in many polysilicon factories related to the production of PV modules, dumping costs, or discharging toxic substances in the environment instead of recycling them (Liu, 2017). However, it should be considered that this happens unfortunately not only for the construction of PV panels but for the construction of boilers and structures for conventional thermal power plants.

As time goes by, and thanks to the development of technology, it will be possible to use less material to cover the same functions and, as schematically shown in Figure 2.9, cover more functions with the same equipment and a fraction of the material needed in earlier technology developments.

Humanity has not yet developed the cultural awareness to make full use of these technological advances, but it is also clear that this will occur in future and will contribute to an increased standard of living and reduce the environmental impact at the same time.

However, today it is impossible or not practical to think of a medical apparatus without non-biodegradable plastic, an airplane without Perspex, or chemical tanks without PVC or rubber lining.

Among various initiatives, the cradle-to-cradle concept was developed to identify the concept of technological nutrient McDonough et al., 2013), which means the complete recycling of

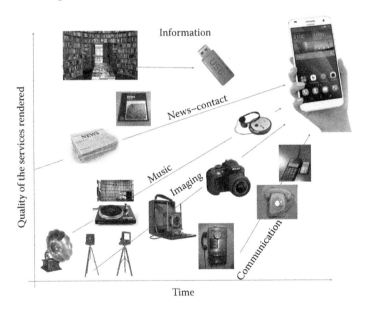

FIGURE 2.9   Technological functions and mass compaction.

technologically non-naturally reversible pollutants, which instead of being wasted to a grave where they mix with biological nutrient and remain forever a pollutant, they are endlessly recycled for new manufacturing exigencies, maintaining their value a prime material source.

Naturally non-reversible pollutants, therefore, can be handled by a combined action aiming at:

1. Reducing their volume and mass

2. Using pollution of the environment as a material source for manufacturing

3. Designing for endless recycle

4. Designing (if possible) for biodegradable use

2.5 TECHNOLOGY AND THE REINSTATEMENT OF A NEW ABUNDANCE

As humanity progresses, it is impossible to let the natural ecosystem upcycling mechanism handle the volumes of waste and pollution generated by anthropogenic activities.

Technologies are gradually being developed with the feature of reducing the amount of waste for a required service and at the same time are capable of upcycling waste as a resource, as indicated in Figure 2.10.

Several alternatives are possible to the flow diagram of Figure 2.5 and will be more likely implementable as technology evolves.

The scheme indicated in Figure 2.10 shows an alternative pattern that would lead to reinstatement of the abundance that was originally in nature and enable the time technosphere to deal with the necessity of current living standards. As can be seen in Figure 2.10, while anthropogenic energy requirements are satisfied by renewable energy, more energy is extracted from the organic energy residue of the waste generated by anthropogenic activities.

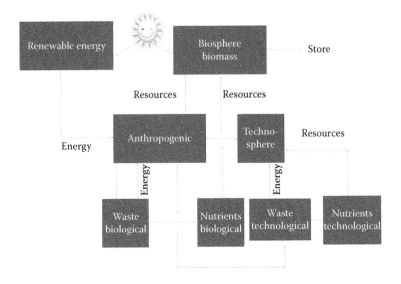

FIGURE 2.10   An abundant flow of exergy and resources in future scenarios.

Waste becomes the source of the material required from the technosphere in a cycle that does not require it to withdraw more resources from the lithosphere for manufacturing process.

Treated organic waste is then returned to the biosphere with all the biological nutrients that are necessary to it. In this cycle, the biosphere keeps renewing itself and storing the sun's energy in the form of more resources at high chemical potential.

It is known, for instance, that many modern agricultural soils are significantly degraded (Gilroy et al., 2008) mainly due to a reduction in nutrient retention and in soil organic matter.

Phosphorus is likely to become a limiting nutrient resource in coming years, and in this respect, the recovery of biosolids has the potential to not only increase organic matter and support organic food yield, but also to build up organic matter in a relatively short time. This equals an increased soil energy, and therefore biosphere energy, and generates again a process of abundance.

However, as indicated later in this book, technology is gradually changing the impact on the environment, and technological developments offer solutions to the process of pollutants removal that were neither before economical nor feasible. These are, for instance, phosphate recovery, plastic recovery and reprocessing, renewable energy that reduces carbon emissions, etc., and it will be possible to use technology to make the planet more sustainable, not less.

In this case, it is possible to define an anthropogenic velocity of pollutant removal that is proportional to the level of technology, as indicated in Formula 2.18:

$$\mu_{antrem} = -K_2 \cdot T^n \tag{2.18}$$

where n is a coefficient that depends on how judiciously environmental activities are managed and how technology can contribute to minimizing the environmental impact generated by anthropogenic activities:

$$V \cdot \frac{d\omega}{dt} = +K \cdot P \cdot A \cdot T - \mu \cdot \omega \cdot V_1 - K_2 \cdot T^n \tag{2.19}$$

This equation states that the accumulation of a pollutant ω in the environment with time is equal to the amount of pollution generated by humanity minus the amount that the biosphere upcycles minus the amount that technology can segregate and reprocess.

If the contribution of technology to mitigate and eliminate the environmental impact is considered, then the sustainability threshold becomes the values stated in Equation 2.20:

$$\frac{K \cdot P \cdot A \cdot T}{\mu \cdot \omega \cdot V_1 + K_2 \cdot T^n} = 1 \tag{2.20}$$

The formula, therefore, allows the possibility to have a growing population, living an affluent lifestyle, supported by a technological

TABLE 2.4   Resources Sustainability Thresholds

Item	Sustainability Index	Criterion
1	$\dfrac{K \cdot P \cdot A \cdot T}{\mu \cdot \omega \cdot V_1} \leq 1$	Naturally reversible pollutants, in nature without human intervention. The threshold of sustainability occurs when the waste discharge rate equals the Earth upcycling rate.
2	$\dfrac{K \cdot P \cdot A \cdot T}{\mu \cdot \omega \cdot V_1 + K_2 \cdot T^n} \leq 1$	Naturally reversible pollutants, with human intervention. The threshold of sustainability occurs when the waste discharge rate equals the Earth upcycling rate plus the human upcycling rate.
3	$\dfrac{K \cdot P \cdot A \cdot T}{K_2 \cdot T^n} \leq 1$	Non-naturally reversible pollutants, with human intervention. The threshold of sustainability equals when the waste discharge equals the human upcycling rate.

level of implementation that would offset their emission in the environment.

It would then be possible that the ratio that defines the sustainability threshold becomes less than 1, and humanity thrives again in a regime of abundance.

Based on the previous considerations, the sustainability indexes are summarized in Table 2.4.

One important conclusion from Table 2.4 is that technology can have a positive impact on humanity sustainability approach, both in supporting the natural pollutant upcycling mechanism but also importantly in the management of non-natural and toxic pollutants.

CHAPTER 3

Technology, Affluence and Environment

A Misconception to Change

OR MANY YEARS, THE industrial development generations have grown with the idea that environmental protection was incompatible with technological progress, human welfare, and population increase. Accordingly, these generations resigned themselves to accept and justify the overall disruption of the planet as a compromise against modern living style and commodities.

In some cases, humanity created natural parks, and these oases, in the midst of an environmental disruption, were supposed to be a reminder of what nature would be like if nature still existed.

While the environment is gradually being destroyed for increased wealth, some environmental and social movements advocate a "happy degrowth" whereby a more frugal lifestyle would enable to mitigate the impact on the environment.

The leading thought in this part of the book is very simple. It does not have to be this way, and humanity does not have to give up anything, neither the beauty of nature nor the comfort of technology. Quite the contrary. Technological and cultural progress may lead us to reverse this, and there can be at the same time an affluent, highly technological society, supporting a large population living in perfect harmony with the environment.

This thought will be again dealt with in the economic part and used to support the concept that economic growth and sustainability are not at odds with each other. Businesses do not have to necessarily destroy the world to make profit. Quite the contrary.

Also in this case, humanity does not have to give up its economy to respect the planet, but profit and purpose can be combined.

3.1 A FORMULA THAT SYMBOLIZES THE PERCEPTION OF FEW GENERATIONS

The impact, population, affluence, and technology (IPAT) equation was developed in the 1970s by Barry Commoner, Paul R. Ehrlich, and John Holdren (Commoner, 1972) and symbolizes the approach towards environmental impact of that period of rapid industrial and technological development.

The formula attempts to describe the environmental impact caused by human activities in terms of three variables: population, levels of consumption (which is indicated in the term "affluence"), and impact per unit of resource use, which is termed "technology."

Equation 3.1 indicates that the environmental impact I is directly proportional to population [P], affluence [A], and technology [T]:

$$I = P \cdot A \cdot T \tag{3.1}$$

where I = environmental impact, P = population, A = affluence, T = technology.

The interpretation of this formula can be in order not to have any environmental impact, humanity should be either disappear from the Earth (population P equals 0), or it should decide to live as in the stone age (with technology equal to zero) or in extreme poverty (affluence 0).

This is a very sad and demotivating approach, as in practical terms, this formula states that humanity has to destroy the world to move on or choose between a "happy degrowth" that has been argued by some environmentalists, or a dooming future of pollution and ecological disruption.

Looking at earlier statistics illustrating the ecological footprint against the human development index, it would appear that the IPAT formula is correct. For instance, Figure 3.1 shows human welfare (in other terms, affluence) and ecological footprint of various societies in the contemporary world.

As can be seen, societies with high technological and affluence levels generate in turn a high ecological footprint (above 5 global hectares per capita), while relatively undeveloped countries have an ecological footprint of less than 2 global hectares per capita.

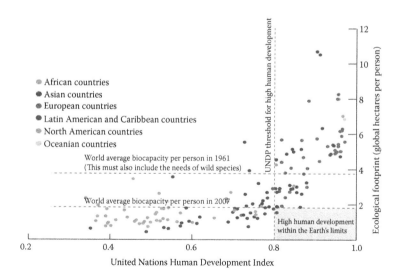

FIGURE 3.1 Schematic comparison between human welfare and ecological footprint. (Adapted from Ewing, B. et al., 2010. *The Ecological Footprint Atlas 2010*. Global Footprint Network: Oakland, CA, p. 29.)

Only a few countries worldwide offer a high human development potential and a low ecological footprint.

The unavoidable result of this approach, suggested by the application of the IPAT formula, appears to be an irreversible process of environmental degradation with the consequence that the only remedy possible to mankind to mitigate the environmental impact and prevent the planet from taking a road to irreversible disruption would be either reducing the population, or the affluence, or a combination of both.

These considerations have been certainly valid for the 1970s period, as this situation was the common pattern to the social economic development of mass consumer goods societies at that time. Furthermore, as can be seen from Figure 2.6, the order of magnitude of the anthropogenic activities and their impact on the environment was considered negligible compared to the huge buffer that was offered at those times by nature.

One limitation of the IPAT formula is that it does not consider the inherent capacity of the Earth, through its biological systems, to metabolize and upcycle natural pollutants and regenerate fresh elements, restoring a situation of environmental balance.

In addition, social behavioral patterns have changed since the 1970s. In this period, mass production and consumerism were the main factors contributing to the increase in affluence and enhanced technology in society. However, with the development of more modern societies, these patterns have radically changed in several situations, and so has the perception of affluence.

As technology develops, the relation of direct proportionality between technology, affluence, and their impact on the environment shows a limited validity. As technology moves forward, sustainable development becomes more closely linked to the use of technology, and the effects of technology are not only negative.

Today, in many cases, it can be observed that at comparable levels of affluence and population, highly technological societies have the ability to support large populations and at the same time generate a lower environmental impact than less developed societies.

While the proportionality of the environmental impact on population and affluence is unquestionably valid, the relation between environmental impact, technology, and affluence today has often appeared to be changing and developing to indirect proportionality, and technological development has often become a strong tool in the hands of humanity to mitigate, or in some cases reverse back, the environmental impact.

Today, technology and sustainability are in an ambivalent relationship. The use of modern technology is on one hand partly responsible for today's problems, but on the other hand, the need for more sustainability depends to a considerable extent on new technical solutions.

This chapter of the book aims to find a new correlation between environmental impact, affluence, and technology that reflects a gradual evolution and changing behavioral pattern whereby it is possible to look without fear to the technological transformation of economy and rely on them to create a new abundance.

3.2 A GRADUALLY CHANGING BEHAVIORAL PATTERN

A modern sustainable society can only function if it can make use of reliable technical systems in an efficient manner and does not impose the concept of sustainability to the media. This is why it is important to create awareness on the sustainability topic and cultivate fundamental criteria for sustainability.

A culture of sustainability can at the same time make use of, but also support, the development of new technologies in the implementation of day-to-day life models for sustainability purposes.

The approach previously used was to observe various societies' reactions to the availability of new technologies and how they applied this to the purpose of reducing environmental impact. Those type of societies who had similar patterns and behavioral characteristics have been observed, and, therefore, it was possible to establish if the impacts they generated in these environments are similar in order to identify communalities and similarities.

On the other hand, assessing a pattern common to a type of society is not so straightforward. Humans live all sorts of different ways and have different social and behavioral backgrounds. Lifestyle affluence and technology interact through more complex multifaceted systems; therefore, establishing an exact correlation between one factor and another is very difficult.

Correlation and association are, therefore, used to describe the relationship between two factors indicating a cause and effect relationship.

One of the first impression is derived from the analysis of the CO_2 emissions from fossil fuels and cement per capita between 1990 to 2000 and between 2000 and 2010, which is shown in Figure 3.2 for countries that in 1990 already had a substantial human development index and, therefore, were in an industrialized condition.

It can be observed that in the early phase (between 1990 and 2000) while societies increased their affluence and their

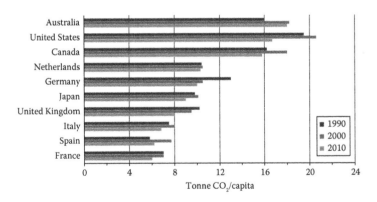

FIGURE 3.2 CO_2 emission from fossil fuel use and cement production per capita for industrialized countries. (Adapted from Olivier, J. G. J. et al., 2012. Trends in global CO_2; emissions 2012 Report, © PBL Netherlands Environmental Assessment Agency, PBL publication number: 500114022.)

technological domain, the environmental impact related to the CO_2 emission per capita has increased somewhat dramatically. This trend is reversed between 2000 and 2010 when CO_2 emissions from fossil fuels and cement per capita decreased to levels, in some cases, lower than in 1990. In this period, the levels of affluence and technology kept increasing, but more conscious environmental regulations and personal behavior enabled a pro capita reduction.

The increase in the CO_2 emissions per capita appears to be a standard pattern that occurs when societies develop and when they pass from poor rural environments to become more technologically developed.

The graph indicated in Figure 3.3 shows the CO_2 emission from fossil fuel and cement per capita for some developing countries.

This is a standard phase in the development of a society where large mass consumer goods are offered to the market on a large scale for a low price. The production is often undertaken in deregulated markets with no attention to ecological footprint.

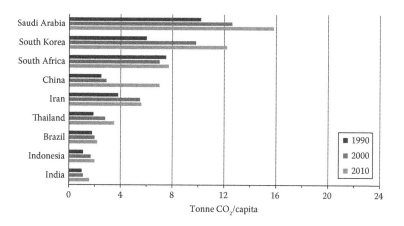

FIGURE 3.3 CO_2 emission from fossil fuel use and cement production per capita. (Adapted from Olivier, J. G. J. et al., 2012. Trends in global CO_2; emissions 2012 Report, © PBL Netherlands Environmental Assessment Agency, PBL publication number: 500114022.)

As can be seen from Figure 3.3, as the countries examined developed from 1990 to 2000 and 2010, increasing and not reversing their affluence and their technological domain, the environmental impact related to the CO_2 emission per capita continued to increase, sometimes dramatically.

As time passed, it has become more evident that the dependence of the environmental impact on affluence and technology is in turn dependent on technology, and it has also become clear that wealthier and more technological societies can generate a lower environmental impact thanks to the help and support of technology and a cultural awareness on sustainability and environmental impact.

Figure 3.4 shows the percentage of waste recycled—waste incinerated and landfilled in various countries in the European Union. As can be seen from Figure 3.4, countries that have developed a higher technology and better economy and affluence have also managed to decrease the environmental impact. For instance, Germany and Sweden have no environmental impact related to waste discharge in landfills, as they recycle and compost the majority of their waste.

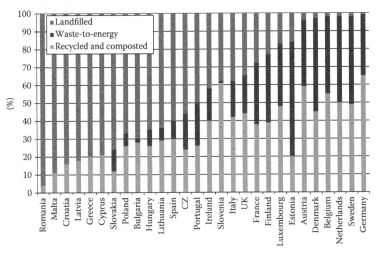

FIGURE 3.4   Percentage of waste recycled—waste incinerated and landfilled against countries. 2013 MSW Treatment in EU28 Eurostat.

These trends can be extrapolated to and associated with the creation and development of more efficient and less expensive renewable energy solutions, as stated in Figure 3.5.

In principle, an infinitely technological society could support a large population with high living standards and at the same time eliminate any environmental impact.

Today the question is asked (Anderson, 2009) whether we could modify the IPAT formula and move T (technology) to the denominator of the IPAT formula (Equation 3.2) so that it becomes:

$$I = \frac{\alpha(\omega)p \cdot A}{T} \tag{3.2}$$

where I is environmental impact, P is population, A is affluence, T is technology, and α is a factor that depends on the cultural and social

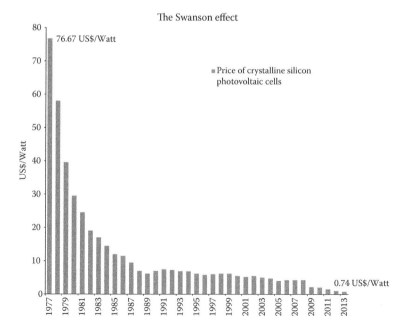

FIGURE 3.5 Silicon PV cells price history. (Bloomberg New Energy and Finance, pv.energy.com.)

development of the society, which is indicated with the parameter ω and therefore changes from place to place in the world.

It is possible based on the previous formula to consider technology and cultural awareness as an ally to the purpose of mitigating the environmental impact rather than an enemy. The leading point of this approach is that if a society was infinitely technological and environmentally conscious, it could support a large population with high living standards without increasing the environmental impact, reducing our impact to zero.

This is happening already. Not only is technological innovation booming, but it is rapidly shifting towards sustainable solutions. For example, many of the World Economic Forum's top 10 most promising technologies have a clear environmental and social focus, such as energy-efficient water purification, enhanced nutrition to drive health at the molecular level, carbon dioxide (CO_2) conversion, precise drug delivery through nanoscale engineering, organic electronics, and photovoltaics (PV).

Figure 3.5 shows the price history of silicon PV cells. In early 1977, per watt cost of these panels was $76.67, over time, this price dropped considerably to $0.36 in 2014. This is the result of policies and of massing investments.

However, this is not only related to the renewable energy resources. In the municipal waste water treatment system, a similar trend can be observed.

Figure 3.6 (O'Callaghan, 2015) shows the cumulative installed phosphorous removal capacity trend.

As can be seen in Figure 3.6, there has been an exponential increase in the struvite technology adoption between 2013 and 2014, with several plants installed mainly in northern Europe and the United States.

Therefore, environmental impact can be negative, and it can be inversely proportional to both affluence and technology. The graph indicated in Figure 3.7 qualitatively describes the relation between affluence (A) and environmental impact (I) for the field of municipal waste water discharge in different societies and

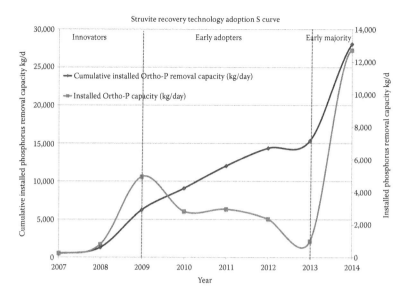

FIGURE 3.6   Cumulative installed phosphorous removal capacity trend. (From Paul O'Callaghan, 2015. Water energy exchange WEX. *Global 2015 Conference Proceedings Water, Energy and the Zero Waste Society*, February 23rd–25th, Istanbul.)

environments. The impact that the technological development of waste water treatment technologies had on both on maintaining the high level of affluence and, therefore, high living standards and controlling or mitigating the related environmental impact compared to relatively non-technological societies becomes evident.

In particular, in areas with poor affluence and no sanitary structures, often in large communities, people there have no toilets or sewage infrastructure. Open defecation is the only option, with resulting hygiene problems and large environmental issues for large populations.

The extreme opposite occurs in highly developed societies where technologies are implemented today that recover energy and nutrients from the process of waste water treatment through adequate treatment.

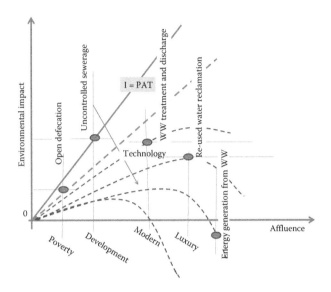

FIGURE 3.7 Relationship between environmental impact affluence at different technological stages for waste water generation.

The practice of energy reclamation from waste water not only provides a negative energy and CO_2 cycle but also no waste to the environment, not only the plant-produced biogas, thus reducing the release of greenhouse gas and producing energy. This may produce high-quality fertilizer that can enrich the soil.

3.3 A NEW FORMULA FOR A FUTURE OF RENEWED ABUNDANCE

Perhaps it will not be possible for quite some time to put the factor T in the denominator, as Ray Anderson suggested. It is also not possible to erase all the negative impact that has been made so far by previous technology use. However, it is necessary to adjust the IPAT formula to reflect future approaches to environmental impact.

It is also necessary to adjust the IPAT formula because, as it was seen in earlier chapters, nature has the capability of upcycling, to some extent, part of humanity's environmental impact.

However, future scenarios and technology development may offer opportunities to improve living standards that foresee reversal of the impact on the environment already generated and reclaim back the polluted areas and elements thanks to technology development.

The full or partial achievement of future renewed abundance could be in reach of future generations and hopefully is an objective of a new wave of economic development.

Therefore, Equation 3.3 may be used to describe the environmental input in this situation and is schematized by the following equation:

$$I = P \cdot A \cdot T - (\mu \cdot \omega \cdot V_1 + A_1 \cdot T^n) \tag{3.3}$$

where

$$A \geq A_1$$

I is the environmental input, T is technology, A and A_1 are affluence factors in the lag industrial phase development, and A_1 is the technological and environmentally conscious society environmental factor.

The term μ represents the inherent planet capacity to revert the environmental impact by upcycling the pollutants ω, and V_1 is the volume of the biosphere, The previous formula states that the environmental impact equals the anthropogenic impact (PAT: population, affluence, and technology) minus the inherent planet upcycling capacity $\mu \cdot \omega \cdot V_1$ minus the anthropogenic technology contribution to reverse the impact $A_1 \cdot T^n$.

In this equation, n is a power factor that depends on how judiciously future generations manage their environmental activities and how technology is able to contribute to minimizing the environmental impact generated by anthropogenic activities.

Both A_1 and n will depend on the technology, the innovation, and the maturity of a society towards the environment.

Figure 3.8 shows the impact of well-developed technologies (as a combination of type of technology, innovation, and maturity)

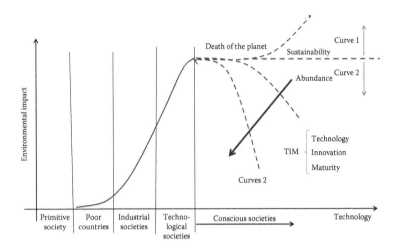

FIGURE 3.8 Environmental impact development against technology development in future scenarios.

on the environment impact and the three possible scenarios that can develop as a result of humanity used of technology.

When technology factor is very low, $A \geq A_1$, the formula remains the same as the one proposed by Barry Commoner, Paul R. Ehrlich, and John Holdren corrected by the upcycling capacity of the planet $\mu \cdot \omega \cdot V_1$. However, as the technology factor increases, the environmental impact increase tends to smooth and, depending on the A_1 factors, the environmental impact (EI) could eventually decrease.

The second term of the previous equation, namely $-A1 \cdot T^n$, can be regarded as the mathematical expression of what can be identified today with the term "sustainability, the new wave of the economy" (Allem, 2016), which means an economy where profit is generated through environment and sustainable economy. The term $P \cdot A \cdot T$ represents the traditional industrial economy and the environmental impact according to the classical theory.

TABLE 3.1   Resources Sustainability Thresholds Modified

Item	Sustainability Index	Criterion
1	$\dfrac{P \cdot A \cdot T}{\mu \cdot \omega \cdot V_1 + A_1 \cdot T^n} \geq 1$	Unsustainable development. The environmental impact generated by humanity is higher than the combined biosphere and technology reversal mechanics.
2	$\dfrac{P \cdot A \cdot T}{\mu \cdot \omega \cdot V_1 + A_1 \cdot T^n} = 1$	Sustainable development. The environmental impact generated by humanity is equal to the combined biosphere and technology reversal mechanics.
3	$\dfrac{P \cdot A \cdot T}{\mu \cdot \omega \cdot V_1 + A_1 \cdot T^n} \leq 1$	Abundance. The Environmental impact generated by humanity is lower that the combined biosphere and technology environmental impact reversal mechanisms.

The sustainability threshold, therefore, can be identified as indicated in Equation 3.4:

$$P \cdot A \cdot T = \mu \cdot \omega \cdot V_1 + A_1 \cdot T^n \qquad (3.4)$$

and the sustainability indexes as indicated in Table 3.1.

As can be seen from Figure 3.8, a scenario where technology will assist future societies in environmental remediation is possible, and this may trigger a reverse environmental impact.

If the environmental impact could decrease and become negative, there would be different development scenarios available in future, and these are illustrated in Figure 3.8, showing possible different trends of the cumulative Earth environmental impact against the time.

The first scenario that will continue the current trend would bring about a continuous increment of the degree of environmental damage made to the planet, as shown in curve 1, until the death of the planet. The second one (curve 2) refers to the maintenance of the status quo, and finally, the last one (curve 3) is the scenario where gradually the planetary environmental damages are recuperated via different degrees of technological means.

3.4 A NEW CONCEPT FOR AFFLUENCE

The definition of affluence is traditionally associated to abundance of material goods, riches, and wealth.

Clearly, the IPAT formula was associating this to an inherent environmental impact related to the energy and material footprint associated to their manufacturing operation and disposal.

A rise in industrialization and economic activities had very strong impact in areas such as job creation and improvement in standard of living. This has triggered the initial association of technology and affluence to quality of life, which is now taking a different direction on the lifestyle model.

Many countries have consolidated their position and have become lucrative locations for investment and setting up of manufacturing centers. The more industries that are attracted, the higher is the investment in the country.

Therefore, somewhere in the past, humanity started identifying the concept of affluence with aspects of the human life that are not necessarily associated to the quality of life.

In fact, in this race for economic supremacy, corners have often been cut and rules flouted.

The resultant factor is pollution resulting from disposal of wastes, discharging effluents into the ground and finding their way to the ground water. The effect of this pollution sometimes manifests quickly, while in other cases, a generation ends up suffering from diseases and health challenges due to genetic disorders (McDonough et al., 2013). This is often the cause of a too rapid and unplanned development that intertwines various aspects related to population growth and poverty.

Today, a significant section of the world population is languishing below poverty lines, and they do not have access to basic living resources; therefore, speaking of evolution of human needs and quality of life may appear unrealistic or even offensive.

As mankind progresses towards new development, the exigencies of life change and evolve towards a more inclusive and responsible living style, which includes a higher responsibility towards others, towards future generations, and towards the planet environment. Accordingly, this brings about a lower dependence on mass consumerism as it is intended today.

Future generations may hopefully aim at shifting the paradigm of their life from the need of ensuring security through material and wealth to a more holistic comprehension of the quality of life where clearly independence, access to plentiful nature, beauty, peace, and environmental sustainability will play a pivotal role.

This shifting paradigm is indicated in Figure 3.9, which—once everybody's basic human rights are granted—is aims to show future generations the pillars of well-being.

Sustainable development in this respect has the enormous potential to generate an economy that touches every aspect of our life, involving both the purely financial aspect but also the economic and, sociopolitical objectives while ensuring better living standards, conserving ecology, the biosphere, and our planet.

When affluence is understood only from the five sensory points of view, it becomes unavoidable to link it to the large material

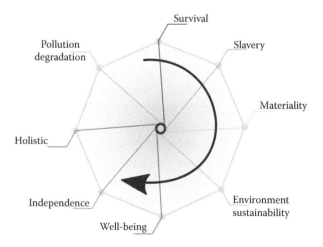

FIGURE 3.9   Evolution of societies and living objectives.

assets that can be produced, acquired, traded, spent, conquered, and stolen.

Humanity is going to grow into a highly functioning society, and we can expect no limits in our ability to solve life's difficulties and transform our lives day-to-day to be longer, healthier, spiritually elevated, and free from material needs. Accordingly, it may well be possible that living in peace with the environment and having access to plentiful natural beauty will become the criteria that determines future generations' affluence.

If affluence can be related to quality of life, it is unavoidable to believe that a change in the relation between society and environment will take place once all other aspects related to adequate access to food, water, education, and financial security have been addressed.

This technological development is aimed at reducing waste, reducing the material and energy requirements, and at the same time ensuring an even better standard of life.

This is gradually done by improving efficiency and consequently reducing the impact on the environment, and it can be the base of a new wave of economic development that would gradually restitute the planet to nature and find a harmonious way of living on Earth for mankind.

There is a lot that can be done in residences, localities, communities, and larger geographic areas to both ensure more civilized living standards and at the same time lower environmental impact, and this can generate a new momentum for the economy, no longer producing large mass consumer goods but a better living quality and more efficient life-sustaining system.

Consequently, well-being is less associated to material possessions and more to availability of time and plentiful access to nature and beauty as well as security.

In a global development scenario, which is previously described, the meaning of economic growth as it is now considered may not be an accurate portrait of the economy.

CHAPTER 4

Create an Economy of Abundance

THE IDEA THAT HUMANITY must become more environmentally friendly and operate in a sustainable manner for philanthropic reasons is purely utopian.

If everyone is conscious and feels responsible for finding a solution to sustainability problems—and this is still far from the truth—the economy of sustainability needs to come to terms with the reality of business. The reality is paradoxical: producing something in a sustainable manner decreases a product competitive edge in today's market, more so if competition comes from countries with relaxed or no environmental regulation in place.

As sad as it sounds, if anyone wishes to find a solution, he must face the reality of business in present times. Preaching about nature conservation and forthcoming dooming scenarios is not going to change the low- to mid-term priorities of the business community.

The economy of a single business or of an individual country or a corporate organization cannot be asked to become sustainable unless everyone does it, and unless there is a common approach to sustainability that brings competition to the same table.

No plan to reverse environmental degradation can be enacted if it requires a wholesale change in the basic dynamics of the market. This includes the fact that costs and growth are the primary contributors to our business choices, and while sustainable solutions can be incentivized, at the same time they cannot be enforced with today's market criteria (Hawken, 1994). The bottom line is that the economy will become sustainable and environmentally friendly if there is profit associated with it.

The leading thought in this part of the book is that there is no contradiction in creating an abundant prosperous technology that preserves the environment and enables us at the same time, to live in perfect harmony with nature.

Growth is one of the main prerogatives of economy as it is perceived today. The economy has to grow in order to be healthy, and for quite a long time, the emphasis has been on how humanity could solve the apparent dichotomy between environmental impact and sustainable economic development, which has been the base of the incompatibility between sustainability and economy.

The dichotomy is between a financial world that demands growth with revenues and profits and environmentalists who prefer lower CO_2 emissions and decreasing population growth, resulting in a lower amount of pollutants volume discharged.

The objective of this part of the book is to demonstrate that this dichotomy does not have reason to exist, and a new wave of economic development can reconcile the exigencies of a healthy economy and an abundantly natural world and that these can coexist.

Beyond humanity's own survival and wellness, there are various driving forces that would lead to a greener economy, and these are going to be analyzed in the following part of the book.

4.1 THE FIRST DRIVING FORCE: EFFICIENCY AND LIFECYCLE OPTIMIZATION

High environmental impacts are often the result of unfortunate economic choices made or promoted upfront by planners on a short-term basis that does not take into account the evolution of the economic scenarios from the time the planning is made.

Many of these economic choices reveal themselves to be inconvenient, if not disastrous, on a long-term basis and through the lifecycle of the initiative.

Some state in the Gulf Corporation Council (GCC), for instance, has massively invested in early 2000s in cogeneration plants that produced water and power from fossil fuels and thermal desalination. The lifecycle of these thermal desalination plants is 20 to 30 years, and the energy footprint of the water produced is in the range of 20–25 kwh/m^3. State-of-the-art technologies at the time were able to produce water from reverse osmosis technologies with an energy footprint of 4 kwh/m^3 with the consequence of millions of U.S. dollars in investments made in technologies that are today obsolete.

The rationale for this choice has been often the low price of energy, but in the lifecycle of the project, the price of energy is escalating, and the technical solution chosen did not consider the entire scenario of operational costs.

Often choosing a more efficient and therefore more environmentally friendly technology is the result of a tradeoff between initial expenditure and long-term savings throughout the lifecycle of the product: saving up front often means spending more in total.

The graph indicated in Figure 4.1 shows how a technical solution that appears more convenient initially, (indicated in the graph with the darker line), may become financially disadvantageous or perhaps even non-sustainable in the long run compared to a technology initially perhaps more expensive but more efficient (indicated in the graph as the lighter line).

This disparity in the lifecycle cost can become even more marked as the cost of energy (indicated in the graph with a dotted line) is going to vary throughout the life of the project and increase year by year, making less energy-efficient solutions even less attractive in the long term. This changes the light line profile in the light dotted

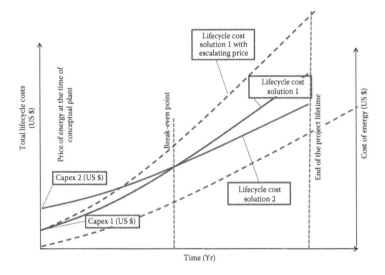

FIGURE 4.1   Technical screening: Lifecycle cost sensitivity against price of energy.

line profile and decreases further the break-even time of a more energy-efficient technology.

The diagram in Figure 4.1 illustrates a situation that has been often occurring in various sectors both related to energy, water, transportation, oil, and gas. This pattern appears particularly marked in geographical areas where energy has been traditionally cheap.

In this situation, low fuel costs have dissuaded additional capital investments that would provide higher efficiency solutions that in turn would reduce fuel consumption and related environmental impacts in favor of initial lower capital expenditure (CAPEX).

Generally, this point of view can be understood when the objective is examined within short time frames. One common misjudgment is to plan development for the future considering today's market conditions.

Since the cost of non-renewable energy sources is unavoidably going to increase, the economical evaluation of traditional fossil fuels versus renewable source needs to be undertaken on a broad timescale and must have a view of inherent social costs that are often not part of the overall evaluation.

The graph indicated in Figure 4.1 shows the historical development of monthly West Texas immediate (WTI) crude oil prices from the year 1948 until 2016.

Despite sharp fluctuations due to mainly important political circumstances, the trend shows a gradual but constant increase in the price of oil (Figure 4.2).

Similar charts can be drawn also for coal and gas prices. Fluctuations in the price often temporarily lead to a decrease due to overproduction or new extraction technologies, but in the long term, the trend always leads to an increase.

With this in mind, planning a technological development in human society such as power projects, a residential building, etc., if taken on a lifecycle holistic view, becomes an extremely delicate process as variables are going to change throughout the life of the project.

On the other hand, the industry often lives on a short cycle, and since it will be dealt with in the next chapter, disregard the

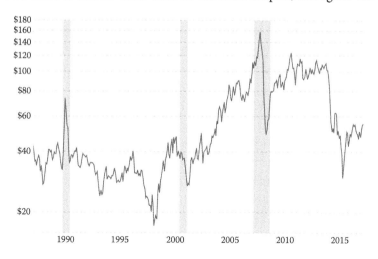

FIGURE 4.2   Crude oil history chart (www.macrotrends.net).

liabilities that the business itself generates during its construction, operation, and dismissal.

4.1.1 Renewable Energy: Gaining Market Share

Renewable energy such as photovoltaic (PV) or wind energy has been subsidized for a long time, and this has underpinned the growth of an industry that is now on par or equally competitive with traditional fossil fuel resources.

On the other hand, technology is developing fast and offers more sustainable and economical solutions.

The graph indicated in Figure 4.3 shows the trends of power tariff from PV and fossil fuels in the Middle East. This graph is derived from the published tariffs of recent major PV and conventional, projects awarded in the area. The trend shows a sharp decline in tariff generation from PV projects, which are now well below the conventional fossil fuel tariffs.

Obviously, each tariff has a different background that depends on many factors such as the history of the project, the country, and its conditions. It is impossible- and it would be wrong- to make hurried conclusions. However, the market trend is very clear, and it becomes even more impressive if one considers that the solar tariffs indicated in Figure 4.3 were not government subsidized, contrary to thermal power where fuel is often provided at subsidized costs.

The latest bid submitted in October 2017 for the 300 MW PV plant in Sakaka, Saudi Arabia is well below the $0.02 threshold at SAR0.0669736/kWh (US$1.786 cents) and represents a global record in power generation.

This tariff also brings about substantial advantages related to the possibility of savings in long, expensive, and energy dissipating transmission lines that convey electricity from the coastline (where the majority of conventional thermal generation assets are located) to the interior of the country where major power consumption occurs.

While today's economy and society cannot depend only on solar energy, it is essential to consider an adequate energy mix that

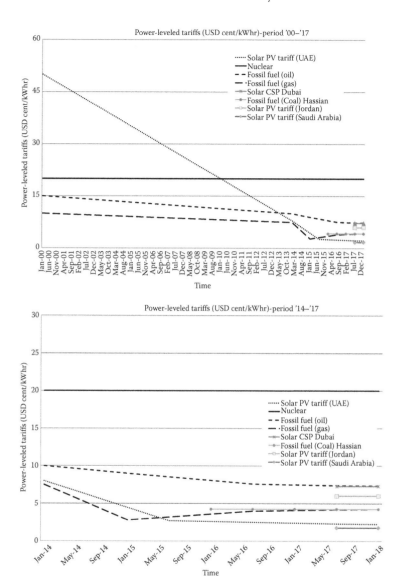

FIGURE 4.3   Power generation tariffs (PV-oil and gas) in GCC. (From MEED 2016, MEED 2017, PVtech 2017.)

covers the energy demand day and night and solve the problems that are posed by the erratic nature of the renewable energy and the need for energy storage. Furthermore, these tariffs were achieved in the Middle East, and it should be considered that the value of the land in a desert area is lower than elsewhere, and irradiations are also higher than in Europe.

On the other hand, the trend shows a decrease in price that is dramatic, and it shows a pattern that is going to continue underpinned by technological developments.

There is another aspect that is often neglected in a fair evaluation of solar tariffs against conventional fossil fuel. A holistic lifecycle view on long-term scenarios shows that the tariffs from conventional fossil fuel are susceptible to severe variations and indexations resulting from fuel price adjustment.

These factors, as indicated in Table 4.1, represent a risk for investors and offtakers for thermal power plants, which become minimal for solar tariffs.

Table 4.1 states that if one invests in a power plant, his costs are going to be affected and indexed by parameters such as fuel cost, chemical and manpower cost escalation, and change in laws of environmental limits. These are going to affect consequentially

TABLE 4.1   Renewable Energy Versus Fossil Fuel Tariff Considerations

Description	Renewable Energy	Fossil Fuel
Sensitivity of power tariff against fuel price	Not applicable	Quite high
Sensitivity of power tariff against manpower-spare and consumable cost	Low	Power plants are highly intensive manned projects and, therefore, quite high
Sensitivity against the risk of new and more stringent environmental criteria	Relatively low	Much higher, as new emission limits for air and water may require retrofit works that would affect the tariff
Impact of soil reclamation and contamination at the end of production time	Negligible	High

his revenue and profit margin in a much more severe way than for a solar plant.

All these aspects are purely commercial and do not consider the social aspects related to the environmental impact and the inherent costs that these social aspects bring about. These constitute the basis for the third driving force of a new economy, as will be indicated in the next chapter.

Social costs are presently non-quantifiable, and this is often the reason why they are minimized or neglected in the whole lifecycle evaluation of the technological options available at the time of decision planning.

However, the fact that these costs are not quantifiable does not imply that they are negligible at all. It implies only that there is no defined party responsible to cater for these costs currently, leaving this liability to public institutions and private citizens.

In the case of energy efficiency, a holistic approach considering the whole costs throughout the lifecycle of the plant would lead to a more energy-efficient and sustainable solution.

In this case, as indicated in Figure 4.4, the financial break-even point is reached after a time whereby it is very difficult to ensure that the financial assumptions still remain valid.

The graph of Figure 4.4 shows a solution that is presenting a lower initial capital investment but an inherent lower efficiency competing with a solution that is more efficient but requires higher initial investments. The less energy-efficient solution is more sensitive against energy cost and in the short time becomes break even with the solution B, which in the long term becomes sharply cheaper.

This generates a conflict between industry, which seeks commercially short-term profitable projects, and public corporations and institutions, which are seeking ideally long-term, sustainable, and viable economic solutions which justify the incentives and support policies that were introduced to develop renewable energy technologies, as described in the following section.

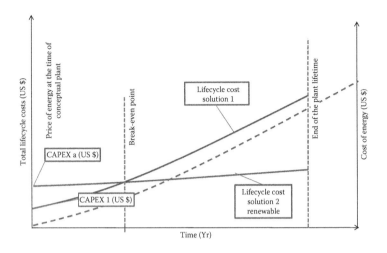

FIGURE 4.4 Technical screening: Lifecycle cost sensitivity against price of energy, sustainable technologies.

4.2 THE SECOND DRIVING FORCE: INCENTIVES AND POLICIES

When everybody has a broad view on the technology options and their consequences in the short and long terms, there would be a consensus to avoid an economy-generating profit at the risk of generating an economic, environmental, and social catastrophe at a later stage.

On the other hand, economy often lives on a short cycle and disregards the liabilities that the business itself generates during its construction, operation, and dismissal, as these are dealt with in later stages of the project life by someone else.

This dichotomy is indicated in Figure 4.5. In many cases, governments and public corporations have been trying to bridge the gap between the short-term and long-term objectives of the economy by providing incentives and subsidies. On the other hand, these solutions are often not considered positively by taxpayers. In this respect, it should be considered that

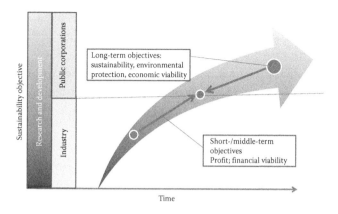

FIGURE 4.5   Public-private objective dichotomy.

incentives and subsidies are generally designed to make viable today a solution that would be most likely in any case viable in the long run.

As it can be seen from the previous Figure 4.5 public corporations, government, policymakers, and visionaries have long-term objectives of creating a sustainable economy, protecting the environment, and avoiding the creation of premises that would, in the long term, generate catastrophic consequences.

On the other hand, the industry operates in accordance to the terms and objectives that are dictated by the economy of competition, and these are short- to middle-term objectives of creating financially viable projects and generating profit.

Governments often dwell in the middle, as the need to make long-term plans often needs to come to terms with the fact that industry objectives are connected to state employment, tax returns, and overall social stability and the need of serving the population with the required infrastructure.

The dichotomy between long-term objectives and short-term financial profitability often appears to generate a gap between the level of affordability of a sustainable solution and its costs, and it is impossible to generate a sustainable practice without an objective consideration to both situations.

Renewable energy solutions such as photovoltaic or wind energy have been subsidized for a long time, and this has underpinned the growth of an industry that is now on par or equally competitive with traditional fossil fuel resources.

Recent growing support from market regulators through tax incentives or grants has resulted in a substantial growth of green bonds. As can be seen in Figure 4.6, according to Moody's, the green bond market is growing exponentially.

The green bonds support the development of specific "green projects" or are issued by companies that invest in the renewable energy sector and, therefore, can benefit from the tax incentives and grants developed by regulators to support the market.

As technology develops, regulators can play a very important role not only through tax breaks and subsidies but also by removing many of the accommodative policies that have been allowed towards the fossil fuel economy which, as described in the next section, do not accurately reflect the social and environmental costs generated by fossil fuel production.

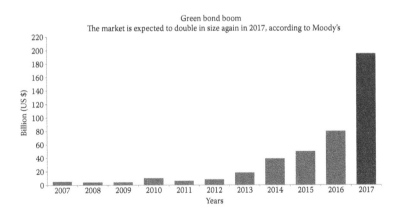

FIGURE 4.6   Green bond market. (Bloomberg, 2017).

4.2.1 Bridging the Gap between Innovation and Industrial Implementation

Sustainability and innovation are two intrinsically connected elements; the scenarios that characterize human existence are continuously evolving, and what is necessary to live sustainably today is not going to be sufficient to live sustainably in the future.

Human requirements develop and may become more or less demanding. If tomorrow's needs are going to be greater than today's, it will be necessary to innovate even more to make sure that these needs are supported sustainably. Accordingly, the same perception of sustainability that humanity has today is going to evolve as humanity progresses towards a better future.

Human acceptance of this aspect of life is also going to change and evolve towards new dimensions, and what is now acceptable is likely to be unacceptable in the future.

It is a matter of life or death to grow. If a society is not growing, it is automatically going to die. Growth involves innovation.

Innovation in turn, cannot be pursued unless it is underpinned by the industry, and generally, the entry barrier of innovative and more sustainable solutions in the market is clashing with higher costs and lower initial reliability.

While new technologies offer the potential to provide a springboard for the future, their commercial viability depends on how successfully all stakeholders are involved, and these include investors, end-users, and governments to create and support opportunities for integrating these technologies on a broader scale.

The overall objective of sustainability and integration of renewable energy sources cannot proceed along without a parallel program of energy efficiency and resource optimization.

While there are several extremely interesting solutions that could, if judiciously developed, offer interesting alternatives to the state of the art water management, there are several risks associated with the adoption of a novel approach.

These risks derive from the relatively short operational experience of the alternative and in society's resistance to their implementation and practices. These constraints presently tend to discourage public utilities from pursuing their use, and in turn, investors often do not see a commercial opportunity in further developing new technologies.

Often a compromise is reached with the introduction of subsidies, and the program of support to innovative solutions commences in the holistic assumption the cost of subsidies is much lower than the costs and consequences of inactions would imply.

Therefore, it is essential to find a way to bridge the gap between research and industry to make sure that new concepts in the overall management of innovative schemes are successful.

4.3 THE THIRD DRIVING FORCE: THE NEED FOR AN ALL-INCLUSIVE ECONOMY

Any rigorous financial model requires that the liabilities accumulated within the lifecycle of the project are paid off within the duration of the project. The economic viability of a business initiative, therefore, means that market operation is sustainable regarding current and projected revenues and that the revenues will be greater than or equal to, all current and planned expenditures within the time of the business. In simple terms, any project or activity that can financially support itself is economically viable. In financial terms, a liability extended to tens or hundreds of years after the project is completed, as it happens today with unsustainable practices, makes no sense. Similarly, someone needs to be clearly allocated to pay for this liability, as it cannot be the whole of humanity.

James Madison in 1789 stated that a federal bond should be repaid within one generation of the debt because "the earth belongs in usufruct to the living. No man can, by natural right, oblige the lands he occupied, or the persons who succeeded him in that occupation, to the payment of debts contracted by him. For if he could, he might, during his own life, eat up the usufruct of lands for several generations to come, and then then the lands would belong to the dead, and not to the living."

If this principle is rigorously applied to the overall economy lifecycle, the assessment of the industrial choices that are made today should be made considering the liabilities that they generate in the time span of few generations.

This concept, coined well before the term sustainability came to industry, is very much in line with Brundtland Report of 1987, which defines sustainable development as the one that meets the needs of the present without compromising the ability of future generations to meet their own needs.

On the other hand, as waste and environmental damage is not fully acknowledged today as a financial liability, and since there is no clarity on to whom and when this liability will be paid, the related financial aspects are neglected and passed on and on.

The assessment of the economic viability of a project is generally based on a financial break-even analysis.

Financial break-even occurs when the project breaks even on a financial basis, that is, when it has a net present value of zero. The earlier the financial break-even occurs, the more the project or the business is profitable.

To determine a project's financial break-even point, the annual operating cash flow that gives it a zero-net present value NPV needs to be established. The formula (Equation 4.1) for the financial break-even quantity is:

$$Q_{financial} = \frac{FC + OCF}{P - VC} \qquad (4.1)$$

and, therefore,

$$Q_{financial} = Q_{cash} + \frac{OCF}{P - VC} \qquad (4.2)$$

where FC = fixed costs; VC = variable cost per unit; P = price per unit; OCF = annual operating cash flow; and Q_{cash} = cash break-even point.

However, in the assessment of the financial viability of a product and of the related lifecycle costs, there are several debts left to be paid to future generations. These are generated by the contamination of soil, water, and air and the associated costs that sooner or later will be required to recover and address this problem. These are also social environmental and sanitary costs related to higher incidence of diseases, long-term social and economic problems whose solution is left to others to be solved.

For instance, nuclear industry accounting did not include in its cost estimates the expense of decommissioning those plants or the thorny, expensive problem of how to store, guard, and protect nuclear waste for a period long into the future, which in the case of plutonium is over 200,000 years.

The single greatest flaw of modern accounting is that the cost and losses of destroying the Earth are absent from the prices in the marketplace. A key piece of information is missing on all levels of the economy, an omission that extends the dominance of industrialism beyond its useful life and prevents a restorative economy from emerging (Hawken, 1994).

If the environmental debt that is generated by these business endeavors was fully taken into account, then the financial viability of some of these business initiatives would most probably resolve for completely different technical solutions and initiatives.

This makes reference, for instance, to the environmental debt that is generated by the construction, operation, and, at the end of the service life, decommissioning of a nuclear or coal-fired power plant, which bring about a debt in terms of air, soil, and water pollution, which is often non-quantifiable in economic terms.

In this case, the correct break-even analysis would be represented by Equation 4.3:

$$Q_{financial} = Q_{cash} + \frac{OCF + ECD}{p - vc - EOD} \qquad (4.3)$$

where

ECD = environmental capital debt, and EOD = environmental operational debt.

The break-even analysis would then take the shape of Figure 4.7 whereby the overall real total cost line would never be able to meet the overall sales line.

On the other hand, because in the definition of economic viability it is not specified when and who is going to pay for this environmental debt and how big in terms of monetary value the debt is going to be when it happens, this debt has always been generally underwritten.

However, as indicated in Figure 4.8, while costs can be hidden, liabilities cannot, and if correctly accounted, certain economic practices and business endeavors would be a complete no-go from a business perspective.

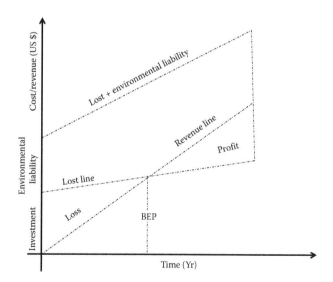

FIGURE 4.7 Financial break-even point adjusted for environmental liability.

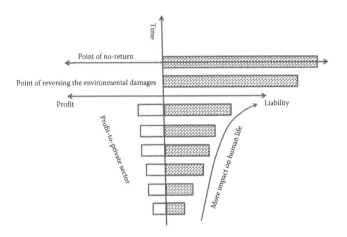

FIGURE 4.8   Profit–liabilities curve schematics.

As indicated in Figure 4.8, the economy has generally made profits for the private sector, generating an environmental debt for future generations. The amount of this environmental debt is often of the same order of magnitude if not higher than the profit that it generated for the economy.

On the other hand, these ethics have been substantially disregarded insofar leading to the gradual accumulation of a greater and greater environmental debt during the current and previous phases of technological development.

As this debt accumulates this would eventually result in a point of no-return.

If this point of no-return is reached, this would lead in turn to an irreversible process of global warming, with consequent melting ice caps, sea levels rising, ruthless floods, droughts, and storms. This will involve a pollution of waterways, oceans, and land that will not be available for agriculture or for carbon and energy storage and the gradual accumulation of toxic waste in landfills, which will have a serious impact on human health and the natural quality of life.

These effects would be accompanied by a contamination of the indoor and outdoor air and a clearly wider gap between the developing and modern worlds.

As indicated in Figure 4.8, it would be logically expected that sometime well before reaching the point of no-return humanity would commit finally to stop and revert the phenomenon.

However, as the environmental liability grows exponentially, it would be necessary at that point in time, to invest a considerably higher amount to revert the situation, correcting the existing system and returning to the point that would enable humanity to continue.

If these activities take the form of an emergency response, for instance, to urban town flooding or to large areas of planet drying, these activities would not generate profit to the society but would simply transform the environmental debt into a financial debt spent in remedy of damages. If the humanity reaction to this situation is to simply curtail consumption, then again, the economy will be shocked as per capita consumption will decrease.

This situation will not lead to extinguishing the environmental debt that was contracted before but will simply deal with the interests that this debt has generated as emergencies, additional health costs, subsidies to a population in distress, etc.

4.3.1 The Need for a Global Approach

One of the problems that makes it most difficult to take a uniform approach from a regulatory standpoint comes from the fact that unsustainable practices and their consequences on the environment are no longer local but have become global.

Greenhouse emissions in one country may have extreme effects somewhere else in the planet thousands of kilometers away and not in the country where these emissions are generated.

A study published in Germany showed that 88% of the conifer forests in eastern and middle Europe are threatened by pollution; 84% of the deciduous forests in eastern Europe are severely damaged. Some of these forests are hundreds of miles from the nearest serious polluter (Hawken, 1994).

As the effect of unsustainable industrial practice is global, setting up governmental policies that enforce the need for payment of the environmental liabilities is useful, but these policies are unlikely to bring about a significant contribution.

In today's market, there are national economies that have taken a courageous step towards sustainability. They have contributed to reducing the world environmental debt with their own portfolio, often at the expense of the competitiveness of their industries in the short to middle terms.

The organic farmer who sells products that do not contain toxins for the consumer, builds up soil quality, sequesters carbon dioxide through its cultivation, and does not use pesticides or herbicides cannot bring goods to market as cheaply as a conventional factory farm using organophosphates and pesticides in today's market.

In this case, when the organic farmer sells his product, he does not sell his environmental credit, while the conventional farmer does not sell the environmental debt associated with his merchandise.

A fair and holistic trade, therefore, should consider that in addition to the products that one country imports and exports there is a corresponding environmental debit or credit that is also created as the goods are manufactured. This requires a global approach.

4.4 SUSTAINABILITY AND ECONOMIC GROWTH: MAKING MONEY RESTORING THE PLANET

There have been several discussions about sustainable economic development and related strategies that enhance environmental quality. These are called sustainable economic development. However, there has never been an analytical approach to how this new green wave of economic development could be compatible with the growth required by today's economy.

The classic theory of economy developed by Solow defines the minimum conditions required to have a growing economy, as dictated by Equation 4.4:

$$\frac{Cap_t}{P_t \cdot T_t} < \left(\frac{s \cdot K}{\alpha + n + \tau} \right)^{\frac{1}{1-\alpha}} \qquad (4.4)$$

where Cap_t is the capital available at the time t; P is the population; T is the level of technological development; K is a constant; s is a saving index; n is the population growth index; τ is the technology development index; and α is the amortization index.

As seen in the previous formula, to have economic growth, the population P_t needs to be high as well the actual level of technological development T_t.

On the contrary, the amortization rate, the technology development index, and the population growth rate decrease the drive towards economic growth.

The combination of the classical economic theories and environmental impact provide a.

In particular, the application of the IPAT formula developed in the 1970s by Barry Commoner, Paul R. Ehrlich, and John Holdren [1] to the equation of Solow's straw and in particular the substitution of the term $P \cdot T = I/A$ into the previous formula results in the following formula that defines the necessary conditions that enable an economic growth:

$$Cap_t < \frac{1}{A}\left(\frac{s \cdot K \cdot I}{\alpha + n + \tau}\right)^{\frac{1}{1-\alpha}} \qquad (4.5)$$

In this formula, I is the environmental impact, and A is the affluence as defined by Commoner and illustrated in the earlier part of this book.

Therefore, according to the classic theory, the generation of environmental impact is the unavoidable requirement to enable the economy to continue growing with the consequent syllogism, whereby economy growth corresponds to Earth disruption. Accordingly, the ecology of commerce stated "business is destroying the world; no one does it better" (Hawken, 1994).

However, the situation could be drastically different if profit can be made and in the meantime the planet's beauty and nature could be restored as described.

The relationship between economic growth, sustainability, and environmental protection can be completely different from the one obtained earlier. If we consider the relationship between environmental impact, affluence, and technology that was elaborated in the previous chapters whereby:

$$I = P \cdot (A \cdot T - A_1 \cdot T^n) \tag{4.6}$$

the relationship between technology and environmental impact can be of indirect proportionality and therefore defined by Equation 4.7:

$$Cap_t < \frac{1}{A} \left(\frac{s \cdot K \cdot f(I)}{\alpha + n + \tau} \right)^{\frac{1}{1-\alpha}} \tag{4.7}$$

where $f(I)$ is a function that can be of direct or indirect proportionality depending on the coefficients n used in the formula defining the environmental impact under the new economy.

Therefore, from a strictly theoretical basis, and without constraints to the free development of economic growth imposed by government or environmental authorities, also from a classic economic analysis, there is no conflict between economic growth and sustainability.

The previous equation shows, in fact, that there is the possibility of an economy where profit is associated with sustainability and environmental protection and, therefore, generates growth in the traditional view of economic development that was indicated. In this case, there is a possibility to achieve a new wave of economic development.

In practice, there will soon be no realistic long-term alternative to humanity except to strive towards a new economic development that combines profit and economic growth with environment, beauty, and sustainability.

Everyone who has consideration to progressive development of the society is aware that new scenarios are opened as humanity evolves and as time passes.

This trend has cyclically occurred in the history of the humanity and has changed gradually for bettering the living style of the society. As time passes, mankind has been able to move forward, and this has enabled the generation of new forms of economy designed around new and more socially acceptable criteria for human life support.

Around this progressive development, humanity should be able to imagine and build a new economy that initially is underpinned by the competitiveness of more efficient products and then develop more holistic exigencies in human life.

As the old economy steps back, a new economy is also likely to develop and generate growth in a gradually more sustainable manner. In this case, a good model is an adaptation to progressive humanity requirement of the nature cycle of production consumption.

As indicated previously, this cycle takes advantage from the energy of the sun and operates in the open metabolism of chemicals for their reproduction. Nature upcycles all byproducts generated to support life of the biosphere to an endless resourcefulness (Stokes, 2016).

Upcycling means eliminating the concept of waste and brings us to a gradual increase of our reserves for future. It should be noted that what we now call reserve was called exergy in the previous chapters of this book.

For reserve it is intended Carbon stored in the soil as organic matter (instead of being oxidized as CO_2), aluminum or plastic stored as technical nutrients, rather than being spread in the environment polluting it. For reserve it is also intended clean land filled with trees and healthy biosphere rather than contaminated by oil as illustrated before. All this implies accumulating resources and storing energy for future use.

Gross domestic product (GDP) can grow underpinned by the consumption of new technologically driven sustainable healthy products, and governments investing in this economy can be directed towards eliminating an ecological liability, transforming it into a credit, and industry investments can be driven by this economy.

Today we look at renewable energy sources as an alternative to fossil fuel for the generation of electricity required for anthropogenic activities, but in future, this energy can be made available also to harvest and store exergy in various manners, for instance in the soil and in living biomass.

At the same time, renewable sources may support humanity to reverse the environmental input created up to this period of industrial development. Most importantly, with abundant energy, there can be abundance of everything else, and a new wave of economic development can begin, a wave where economy can be driven by the generation of better human conditions and, as indicated in Figure 4.9, reproduces on a technological level the generation usage production and storage scheme that occurs in nature.

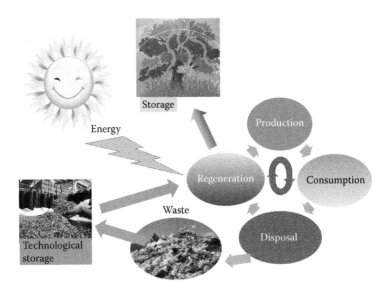

FIGURE 4.9 Technological generation usage upcycling and storage scheme.

4.4.1 Happy Progression: Not Happy Degrowth

There have been various theories for happy degrowth through a more amateurish economy (Nørgård, 2013) aimed at reducing consumption to mitigate the environmental pressure.

Degrowth proposes the concept of enhancing the ecological conditions of the planet by downscaling production and consumption and maintaining a soberer living standard.

This obviously has some merits, since present consumerism schemes have required material possessions as a primary element in defining happiness and well-being.

Clearly, all theories supporting a more sustainable and environmentally friendly living give a great contribution to the goal of achieving an abundant world.

However, apart from the difficulties that this concept would bring about in terms of cultural and economic acceptance and implementation in society, "happy degrowth" offers only a temporary solution and not a radical solution to the sustainability issue.

It is possible to explain how ethically and morally important it is to preserve the planet; however, until this task has an economic return, it will not achieve any substantial progress. The only effective tool against the environment impact is not the return of the Stone Age but the improvement of technology.

Furthermore, "happy degrowth" is not an ambitious enough vision. Degrowth does not conceive and capture the concept of abundance that this planet has been endowed with. It is rather a vision of material misery, frugality, local economy, and the spiritual growth that is allegedly associated with this degrowth that has historically proven to be incompatible with the reality when resources are decreased.

In addition, while in happy degrowth, sometimes the economy shrinks to a minimum level to maintain subsistence; it will not deal with the waste that is already in the ecosystem that continues to contaminate and unavoidably needs to be removed.

This waste, unless a technological shift towards reverting the planet back and reabsorbing it as a technological nutrient occurs, will remain in the planet and will keep contaminating it.

Rather than a happy regression, it would be more beneficial for humanity to strive towards a happy and natural progression, a path where GDP is contributed by elements that are in harmony with the nature of life. This includes nature; tourism; sustainable, organic, and healthy food; heath care; higher education; and happiness with an industry centered along these concepts and making profit.

Consumption does not necessarily mean material good consumption; here, the shift that technology development offers is important.

GDP may be based on production and consumption of sustainable—renewable energy, biological food, and conversion of unsustainable industrial practices to clean ones. GDP may be driven by the conversion of old, ugly, unsustainable residences into beautiful livable cities where clean energy drives tourism, a high level of education, health care, beauty, and contact with nature, a more abundant nature. This nature will be a key element in the economic assessment. This is not degrowth; this is sustainable progress, and it underpins growth and GDP.

4.4.2 Transforming an Environmental Debt into an Environmental Credit

If economy accepts that responsibility for the consequences of business decisions upon human well-being and natural ecosystems, new business can be built up around human well-being requirements, and it will start the process of transforming the environmental debt that has been generated into an environmental credit.

This will in turn revert the liabilities humanity has accrued with the environment and nature into an asset, as indicated in Figure 4.10.

As can be seen in Figure 4.10, the economy is going to invest the effort in the process of transforming the environmental debt

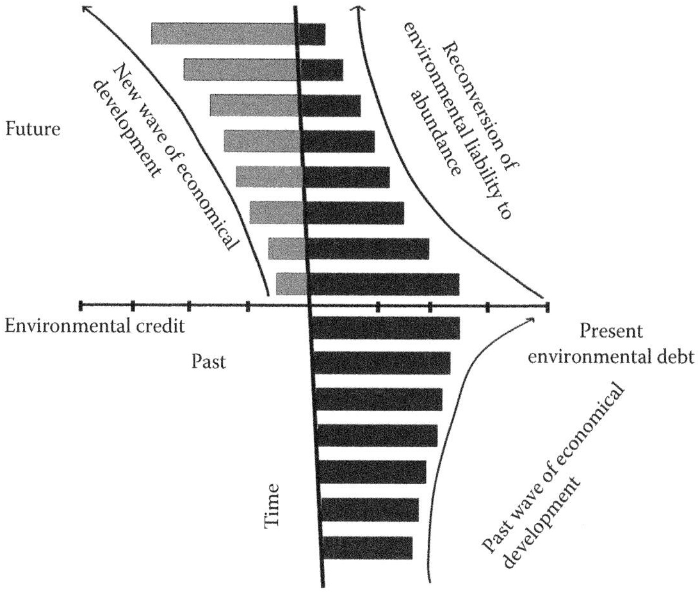

FIGURE 4.10   Transforming the debt into a credit.

into a credit than in paying the debt in one shot later, and this economy in the meantime generates new jobs, new opportunities, and increased wealth.

This obviously would become the driver of a new economy, the new wave of economic development whereby profit is made on purposeful business initiatives.

From a financial point of view, this is not intended to curtail growth and per-capita consumption to maintain the environmental status but simply includes in the per-capita consumption and new parameters that are going to be produced by the new wave of economic developments.

This is happening already. The 2012 Global Green R&D Report found that private investments in clean technology and green economic and commercial solutions reached $3.6tn for the period 2007–2012. This included more than $2tn in renewable energy, $700bn in green construction, $241bn in green R&D,

TABLE 4.2   German Energy Transition Plan

Installed Power Generation	2013	2025
Nuclear	12 GW	0
Coal/Lignite	47 GW	34 GW
Gas	27 GW	30 GW
Wind on-offshore	34 GW	74 GW
PV	37 GW	55 GW
Other	25 GW	25 GW
Conventional	101 GW	77 GW
Renewable	81 GW	141 GW
Total	182 GW	218 GW

$238bn in the smart grid, and $231bn in energy efficiency (Visser, 2014).

The following Table 4.2 shows for the Energy Transition Plan planning and the development of the power generation between 2013 and 2025 as developed by the German grid development plan (Netzentwicklungsplan NEP).

It is possible to easily imagine the implication that the decommissioning of 12 GW nuclear power and 13 GW fossil fuels brings about on the economy, as well as the construction of 40 and 18 GW of wind and photovoltaic power, respectively.

It is perhaps not easy to imagine the fall back on the whole aspect of the economy that will benefit from 2025 onwards in terms of reduced decommissioning operation and maintenance (O&M) costs, health and sanitary costs, and unindexed energy prices. All money that could be invested in other sectors of the economy creates a positive chain that disrupts a negative cycle.

CHAPTER 5

Water

The Path to Sustainability and Abundance

THIS BOOK COULD NOT finish without some consideration to water, the most essential aspect of the biosphere.

Water is a key element in the overall sustainability process, as water is the only element that is necessary to allow the growth of vegetation and forestry, which are elements so crucial not only in terms of greenhouse emissions but also in terms of exergy enrichment of the planet.

As the times goes by and as world population increases, as indicated in Figure 5.1, the available renewable water per capita is sharply decreasing.

This obviously has an important impact, not only for the basic satisfaction of human survival needs but also on the ecosystem and biosphere that, with a limited supply of water, are going to be drastically altered.

Along with the decrease of renewable water, however, there is a constant increase of waste water, which is now seen as a liability to get rid of at the lowest possible cost. In the same way that waste

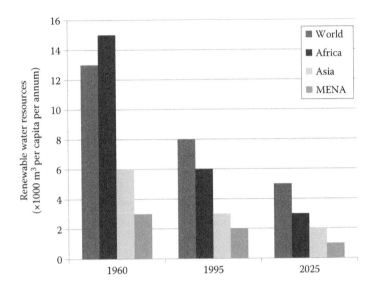

FIGURE 5.1   Annual water availability per capita per year. Derived from http://data.worldbank.org/indicator/ER.H2O.INTR.PC.

needs to become the technological and biological nutrient store for the future, waste water also needs to become the new water source for the future.

This imposes the need for increasing waste water reuse and reclamation. If the concept of cradle-to-cradle is applied to solid waste, then in the case of water, this is an essential requirement for a sustainable future.

Nature upcycles water by phase change with a great dissipation of energy through evaporation condensation and precipitation.

The prevailing pattern for the use of water today follows in principle the scheme that is indicated in Figure 5.2. Water is abstracted from the biosphere, conveyed, treated, and sent for industrial and anthropogenic use. The waste water is either sent back to the biosphere as waste, or it is treated to a minimum quality required to enable the discharge.

With the pattern indicated in Figure 5.2, two main problems occur. The first problem is that the rate of abstraction is gradually

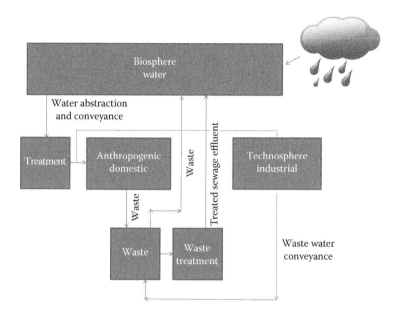

FIGURE 5.2   Current water abstraction and treatment pattern.

becoming higher than the natural rate of replenishment provided by the natural meteorological phenomena; the second one is that the rate of waste water or treated water discharge in the biosphere is gradually becoming higher than the biosphere upcycling capacity. Therefore, while water becomes scarcer, it also becomes more polluted.

Technology now offers the possibility to upcycle water through advance water treatment systems. These processes are industrially well-proven, can be energy negative (and can produce rather than require energy from waste water), and can further produce energy by enhancing the photosynthetic process.

Energy recovery from waste water obviously requires a more capital expensive solution and, therefore, requires an initial higher capital expenditure. However, in markets and regions that have not been adopting accommodating policies towards fossil fuels, these investments generally break even in a very short time without the need of subsidies.

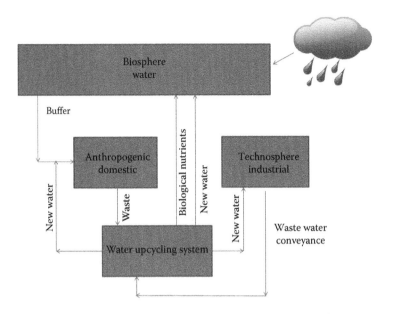

FIGURE 5.3   An abundant flow of water.

The patterns indicated in Figure 5.2 show how an abundant flow of water can be ensured in future scenarios in a similar way as abundance of resources and energy has been treated in the previous chapters of this book.

According to the scheme indicated in Figure 5.3, water abstraction from the biosphere is minimized and left governed by the natural replenishment phenomena. This is achieved by upcycling the water consumed in the technosphere and by domestic anthropogenic uses and generating new water. The sludge currently sent to landfill returns and enriches the biosphere as biological nutrients. As biosphere abstraction is minimized, nature can replenish the natural water resources, leading back to an abundant cycle.

In arid countries, water is generated by seawater desalination. Desalination has been traditionally an energy intensive technology; however, thanks to recent developments, the energy intensity has drastically decreased. Furthermore, in arid countries, several solar projects have been installed with extremely competitive energy

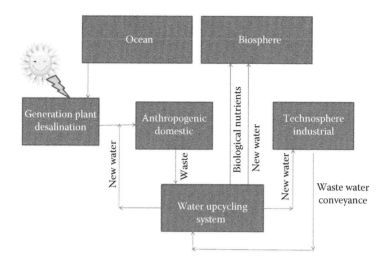

FIGURE 5.4   An abundant flow of water-arid countries.

tariffs, and, therefore, energy-to-desalination projects can be provided by renewable sources.

Figure 5.4 shows how the combination of renewable desalination and water upcycling systems that are energy neutral or energy negative can satisfy water requirements but also contribute to extend or magnify the biosphere. The abundance in this case would means increasing the energy of the biosphere; in other words, turning the desert green.

Disposing waste water without reinserting it into the circle of life brings about at least five associated wastes:

- The life driving and the energy storage possibility that water has for the biosphere.

- The energy associated with the waste water content.

- The nutrients in the waste water.

- The possibility of preserving a natural ecosystem.

- The beauty and wellness that abundant water offers to humanity.

Whenever water can be reclaimed, recycled, treated, and reused for any application that is not only related to domestic or industrial use but also for agriculture, landscaping, and forestation, the planet will receive one of the most significant contributions to its abundance.

For this reason, water resources are essential; they need to be preserved, maintained, and augmented for the health of the planet. In other words, water needs to be used for what has been created by nature: to sustain life.

In this respect, humanity must think to use and reuse water not only for domestic, industrial, and agricultural purposes but also to preserve or reinstate the ecosystem.

5.1 CAPTIVITY VERSUS CENTRALIZATION

In a strive towards sustainability, captive generation of new water with decentralized and captive waste water systems offer the chance of minimizing the displacement of water from the ecosystem and enable the generation of more new water to create more biosphere.

With centralized waste water treatments, the energy footprint related to transmission and distribution constitutes a considerable amount in the overall footprint in the water balance. As a consequence, these systems often do not enable the reclamation of the treated water and related sludge.

Captive generation, in fact, offers the possibility of integrating a water supplyand-demand model that forecasts the differentiation of water use and promotes the adoption of reuse water reclamation for non-domestic purposes or new water for unrestricted use.

The advantage is that this pattern does not abstract large water resources from the environment, leaving the natural phenomena to refill water and water to fulfill the essential task it has in the biosphere: to support life in the biosphere.

In addition, this offers the possibility of reducing the dependence on large water transmission and distribution schemes that often represent the reason for large power consumptions and operation and maintenance (O&M) failures.

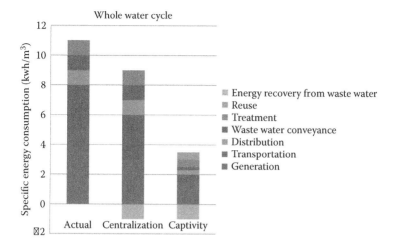

FIGURE 5.5   Captive versus centralized water generation and treatment in the Middle East.

Figure 5.5 shows the comparison of different specific power consumption in different scenarios. The first configuration on the left-hand side of the diagram reflects a typical situation in today's Middle East water schemes.

In arid countries potable water is generated from state-of-the-art seawater reverse osmosis systems at 5 kWh/m³. It requires between 2 to 10 kWh/m³ for transmission and distribution to the end user point. In addition, the waste water produced is conveyed to conventional activated sludge systems, which requires further energy. Although the percentage of water reclamation is increasing, only a small part of waste water is currently reclaimed to a full extent.

The columns on the right-hand side of the graph show the pattern whereby energy consumption is reduced both because of reduction of the flow required from desalination and a shortening of the conveyance energy requirement.

The energy decrease derives from both technology developments and from the fact that more water can be reclaimed from the waste water rather than being generated from desalination.

In addition, energy is generated by anaerobic digestion of waste.

In the future, 60 to 70% of the world's population is going to live in megacities, and these cities must be designed for efficiency, sustainability and beauty. Large centralized water abstraction from dams and centralized waste water treatment schemes with associated large transmission conveyance and distribution schemes are going to offer less sustainable and environmentally friendly and more expensive solutions compared to a decentralized system operating at a district level. Following today's pattern would imply destroying entire ecosystems with dams and long abstraction lines to support consumption in megacities with few recycling opportunities.

On the contrary, captive generation would offer the opportunity to generate and store more water and enrich the biosphere accordingly.

Conclusions

THE LEADING CONCEPT OF this book is abundance: Abundance of energy as a primary source and as a driving force for all life support systems downstream, including the abundance of resources for an affluent life that respects nature. In the absence of sensible anthropogenic activities, the biosphere converts the sun's energy to biochemical energy in excess to what is required to support life. It stores the abundance and increases the Earth's energy levels.

Abundance as a concept beyond sustainability means the enrichment of the planet's energy and resources compared to the status quo.

To support sustainable development and achieve the following target of generating abundance for the next generation, there must be an awareness of nature and the environment that flows down to everyone.

Catastrophic scenarios of drought, famine, pestilence, temperature soaring, and ice melting can be a sad and indeed realistic consequence of our action in the future. Evoking them to support sustainable development will certainly generate concern and perhaps fear, but it will not help too much in creating awareness.

Only awareness, commitment, and love for the environment can create a paradigm shift towards a sustainable and abundant future.

In a similar manner, to boast our indignation, albeit rightful, for the senseless course of action that industry has taken so far in destroying the planet doesn't help as much as showing that a logical alternative path to this course exists. This is a path of abundance that can be chosen for humanity.

This book, I hope, provides a path to retrieve the abundance we have been endowed with, but the conclusions are still to be written.

These will depend on how judiciously we manage the pristine abundance of resources of the planet and keep our minds and hearts open to evolve, continuously looking at the future to restore and maintain that abundance.

Abundance is a way of living that does not only affect our choices towards the world's nature resources but also affects the way we look at our economic development.

When we find balance and we advance confidently in pursuing choices that are aimed at maintaining or increasing that balance in life and in business, then we find a sustainable solution, and we find also a key to prosperity, abundance, happiness, and a positive opening toward the future.

A path that is not at odds with today's understanding of economy is not at odds with our society's rightful desire to evolve technologically and live a more affluent and comfortable life.

This is the path to abundance that this book humbly aims to illustrate and hopes to have provided some useful tools.

References

Allem, R. 2016. *Sustainability: The Fourth Wave of Economy.* ISBN: 978-81-933022-0-0.

Amneus, D. and P. Hawken. 1999. *Natural Capitalism: Creating the Next Industrial Revolution.* ISBN 9780061252792.

Anderson, Ray, at TED2009. 2009. The business logic of sustainability. https://www.ted.com/talks/ray_anderson_on_the_business_logic_of_sustainability/transcript

Bastianoni, S. et al. 1997. Energy/exergy ratio as a measure of the level of organization of systems. *Ecological Modelling.* 99: 33–40.

Beddoe, R. 2008. Overcoming systemic roadblocks to sustainability: The evolutionary redesign of worldviews, institutions, and technologies. *Proceedings of the National Academy of Sciences of the United States of America.* 106: 2483–2489. doi:10.1073/pnas.0812570106.

Bossel, H. 1999. Indicators for sustainable development: Theory, method, applications, ISBN 1-895536-13-8. International Institute for Sustainable Development, Winnipeg, Manitoba, Canada.

Capilla, V. and A. V. Delgado. 2014. *Thanatia: The Destiny of the Earth's Mineral Resources Thermodynamic Cradle-to-Cradle Assessment.* ISBN 978-981-4273-93-0.

Capra, F. 1996. *The Web of Life, Anchor Books.* New York: Random House.

Carnot, S. 1824. Réflexions sur la puissance motrice du feu et sur les machines propres á développer cette puissance (1824) chez Bachelier Libraire quai des augustine. n. 55.

Chen, G. Q. 2005. Exergy consumption of the Earth. *Ecological Modelling.* 184: 363–380.

Cheng, I. et al. 2017. The Great Water Grab: How the Coal Industry is Deepening the Global Water Crisis. Greenpeace International. http://www.greenpeace.org/international/en/publications/Campaign-reports/Climate-Reports/The-Great-Water-Grab/

Chibuike, G. U. et al. 2014. Heavy metal polluted soils: Effect on plants and bioremediation methods. *Applied and Environmental Soil Science.* 2014(2014): Article ID 752708.

Ciais, P. et al. 2013. *Carbon and other biogeochemical cycles in Climate change 2013: The physical science basis. Contribution of Working Group I to the Fifth Assessment Report of the Intergovernmental Panel on Climate Change.* Cambridge University Press, 465–570.

Commoner, B. 1972. The environmental cost of economic growth. In *Population, Resources and the Environment.* Government Printing Office: Washington, DC, pp. 339–363.

Daly, H. 1996. *Beyond Growth: The Economics of Sustainable Development.* ISBN-13: 978-0807047095.

Doane, D. and A. MacGillivray. 2001. Economic sustainability. The business of staying in business. *The SIGMA Project – New Economics Foundation.*

Dyer, W. 2005. *The Power of Intention: Change the Way You Look at Things and the Things You Look at Will Change.* Hay House UK Limited, 2004 ISBN 1781803773, 9781781803776.

EUROSTAT. 2013. EUROSTAT Press Office, Luxembourg. "http://file.scirp.org/Html/10-6702360_48321.htm" \l "return4"

Ewing, B., D. Moore, S. Goldfinger, A. Oursler, A. Reed, and M. Wackernagel. 2010. *The Ecological Footprint Atlas 2010.* Global Footprint Network: Oakland, CA.

Fischer, R. 2015. Simulating carbon stocks and fluxes of an African tropical montane forest with an individual-based forest mode. 10.1371/journal.pone.0123300 April 27, 2015.

German grid development plan [Netzentwicklungsplan NEP].

Gibbs, G. 1873. Graphical methods in the thermodynamics of fluids, transaction of the Connecticut Academy, II, pp. 309–342.

Gilroy, J. J. et al. 2008. Could soil degradation contribute to farmland bird declines? Links between soil penetrability and the abundance of yellow wagtails Motacilla flava in arable fields. *Biological Conservation.* 141: 3116–3126.

Global Footprint Network. 2013. *The National Footprint Accounts, 2012 edition.* Global Footprint Network: Oakland, CA.

Graves, L. 2016, November 16. UAE set for low-cost power benchmark with Hassyan plant. The National.

Greenpeace International Publication. 2016, March 22. The great water grab: How the coal industry is deepening the global water crisis.

Harde, H. 2017. Scrutinizing the carbon cycle and CO_2 residence time in the atmosphere. *Global and Planetary Change.* 152: 19–26.

Hawken, P. 1994. The Ecology of Commerce 0887307043, 9780887307041.

Jepson, E. and A. Haines. (Summer 2003). Under sustainability. *Economic Development Journal.* 2(3): 45–54. http://www.macrotrends. net/1369/crude-oil-price-history-chart.

Kennings, T. 2017. Saudi solar bids debunked: What we know so far. https://www.pv-tech.org/editors-blog/saudi-solar-bids-debunked-what-we-know-so-far.

Lazarus, E., G. Zokai, M. Borucke, D. Panda, K. Iha, J. C. Morales, M. Wackernagel, A. Galli, and N. Gupta. 2014. Working guidebook to the National Footprint Accounts: 2014 Edition. Oakland, CA: Global Footprint Network. Footprint network (June 2014) Working Guidebook to the National Footprint Accounts 2014.

Lippke B. et al. 2011. Life cycle impacts of forest management and wood utilization on carbon mitigation: Knowns and unknowns. *Carbon Management.* 2(3): 303–333.

Liu, Y. 2017. The dirty side of a "green" industry. www.worldwatch.org/ node/5650.

Ingerson, A. L. Global Footprint Network. 2007. *U.S. Forest Carbon and Climate Change.* The Wilderness Society: Washington, DC.

McDonough, W. and M. Braungart. 2002. Cradle to cradle: Remaking the way we make things. *North Point Press.* 193. ISBN 0-86547-587-3.

McDonough, W., M. Braungart, and B. Clinton. 2013. *The Upcycle: Beyond Sustainability—Designing for Abundance.* ISBN-13 9780865477483.

Monod, J. 1949. The growth of bacterial cultures. *Annual Review of Microbiology.* 3: 371–394. (Volume publication date October 1949).

Mullikin, L. 2015. Bagging the anti-baggers. https://delunula.com/ baggin-the-anti-baggers/#.

Murad, S., A. Badran, E. Baydoun, and N. Daghir. 2017. *Water, Energy, and Food Sustainability in the Middle East—The Sustainability Triangle.* Springer.

Nørgård, J. S. 2013. Happy degrowth through more amateur economy. *Journal of Cleaner Production.* 38: 61–70.

O'Callaghan, P. 2015. Water energy exchange WEX. *Global 2015 Conference Proceedings Water, Energy and the Zero Waste Society,* February 23rd–25th, Istanbul.

Olson, J. S. 1963. Energy storage and the balance of producers and decomposers in ecological systems. *Ecology,* 44(2), 322–331.

Olivier, J. G. J. et al. 2011. Trends in global CO_2; emissions 2012 Report, © PBL Netherlands Environmental Assessment Agency, PBL publication number: 500114022.

Patzek, T. W. 2004. Thermodynamics of the corn-ethanol biofuel cycle. *Critical Reviews in Plant Sciences*. 23(6): 519–567. An updated web version is at http://petroleum.berkeley.edu/papers/-patzek/CRPS416-Patzek-Web.pdf.

Post, W. M., A. W. King, and S. D. Wullschleger. 1997. Historical variations in terrestrial biospheric carbon storage. *Global Biogeochemical Cycles*. 11(1): 99–109. doi:10.1029/96GB03942.

Quarati, P. 2016. Negentropy in many-body quantum systems. *Entropy*, 18, 63. Entropy, 18(4), 125.

Ramsayer, K. 2016. ABoVE looks below the surface for carbon answers. Nasa Earth Expedition. https://blogs.nasa.gov/earthexpeditions/2016/07/11/above-looks-below-the-surface-for-carbon-answers/

Rant, Z. 1956. Exergy, a new word for "technical available work. *Forschung auf dem Gebiete des Ingenieurwesens*, 22: 36–37.

Schrödinger, E. 1945. *What is Life? The Physical Aspect of the Living Cell.* Cambridge University Press: Cambridge, UK. ISBN 0-521-42708-8.

Shah, V. 2016. NGO and PepsiCo feud over deforestation, labour claims. http://www.eco-business.com/news/ngo-and-pepsico-feud-over-deforestation-labour-claims/.

Shahan, Z. 2014. http://www.renewableenergyworld.com/ugc/articles/2014/11/5-reasons-solar-power-will-dominate-energy-in-the-next-century.htm

Shizas, I. et al. 2004. Experimental determination of energy content of municipal wastewater. *Journal of Energy Engineering*. 130(2): 45–53.

Smil, V. 2003. *The Earth's Biosphere: Evolution, Dynamics, and Change.* MIT Press. ISBN-10: 0262692988.

Smil, V. 2017. *Energy and Civilization*. ISBN: 9780262035774.

Solow, R. M. 1956. A contribution to the theory of economic growth. *Quarterly Journal of Economics*. Oxford Journals. 70(1): 65–94. doi:10.2307/1884513.

Stokes, K. M. 2016. *Man and the Biosphere: Toward a Coevolutionary Political Economy*. Routledge. ISBN 1315487039, 1315487039.

Susani, L. et al. 2006. Comparison between technological and ecological exergy. *Ecological Modelling*. 193: 447–456.

Swan, T. W. 1956. Economic growth and capital accumulation. *Economic Record. Wiley*. 32(2): 334–361.

Szargut, J. T. 2001. Exergy Analysis of thermal processes and systems with ecological applications: Encyclopedia of life support systems, EOLSS; Theory and Practices for Energy Education, Training, Regulation, and Standards.

Szargut, J. T. 2003. Anthropogenic and natural exergy losses (exergy balance of the Earth's surface and atmosphere. *Energy.* 28(11): 1047–1054.

Tranvik, L. J. et al. 2009. Lakes and reservoirs as regulators of carbon cycling and climate. *Limnology and Oceanography.* 54: 2298–2314.

Tzanakakis, V. A. and A. N. Angelakis. 2011. Chemical exergy as a unified and objective indicator in the assessment and optimization of land treatment systems. *Ecological Modelling.* 222: 3082–3091.

United Nations. 1987. Our common future - Brundtland report. Oxford University Press, ISBN 019282080X.

Visser, W. 2014. How to use technology to make our planet more sustainable, not less. https://www.theguardian.com/sustainable-business/technological-innovation-sustainability-energy-green-investment.

Wall, G. 2011. Tools for sustainable energy engineering: World renewable energy congress.

Wall, G. et al. 2001. On exergy and sustainable development—part 1: Conditions and concepts. *Exergy, An International Journal.* 1(3): 128–145.

Zhu, X.-G. et al. 2008. What is the maximum efficiency with which photosynthesis can convert solar energy into biomass? *Current Opinion in Biotechnology.* Photo retrieved from: www.ombwatch.org

Index